Practical Veterinary Dermatopathology
for the Small Animal Clinician

Sonya V. Bettenay, BVSc
Dip. Ed, MACVSc, FACVSc

CSU Diagnostic Laboratory
Dermatopathology Service
Department of Clinical Sciences
Colorado State University
Fort Collins, CO

Ann M. Hargis, DVM, MS
Diplomate, ACVP

DermatoDiagnostics, Edmonds, WA
Department of Comparative Medicine
University of Washington, Seattle, WA
Phoenix Central Laboratory
Everett, WA

Teton NewMedia

Jackson, Wyoming
www.veterinarywire.com

Executive Editor: Carroll C. Cann
Development Editor: Susan L. Hunsberger
Editor: Nicol Giandomenico
Design and Layout: 5640 Design, fiftysixforty.com
Illustration: Anne Rains

TetonNew Media
P.O. Box 4833
90 East Simpson, Suite 110
Jackson, WY 83001

1-888-770-3165
tetonnewmedia.com

ISBN# 1-893441-96-2

Print Number 5 4 3 2 1

Library of Congress Cataloging-in-Publication Data
Bettenay, Sonya V.
  Practical veterinary dermatopathology for the small animal clinician/Sonya V. Bettenay, Ann M. Hargis.
    p. ; cm.
Includes bibliographical references and index.
ISBN 1-893441-96-2 (alk.paper)
  1. Dogs--Diseases--Diagnosis. 2. Cats--Diseases--Diagnosis. 3. Skin--Diseases--Diagnosis. 4. Veterinary dermatology. 5. Dogs--Histopathology. 6. Cats--Histopathology. 7. Skin--Histopathology. 8. Veterinary histopathology. I. Hargis, Ann M. II. Title.
  [DNLM: 1. Pathology, Veterinary. 2. Biopsy--veterinary. 3. Laboratory Techniques and Procedures--veterinary. 4. Skin Neoplasms--veterinary. SF 769 B565p 2002]
SF992.S55 B47 2002
636.7'08965--DC21

Table of Contents

Acknowledgements

I dedicate this book to my parents, Gwen and Ron, and their partners Bill and Lois. They gave me the early opportunities in life and encouragement that allowed me to pursue my career.

Practical Veterinary Dermatopathology is an attempt to make what I have learned in dermatopathology over the past decade available in a format for veterinarians to use on a daily basis. I hope this book enables a better understanding of the underlying basis of dermatologic disease, special techniques in biopsy sampling and interpretation of dermatopathology reports, and so ultimately benefits patients.

Many people have supported and influenced me throughout my dermatology career thank-you to all of you, but especially Peter Ihrke. With specific regard to this book I would like to thank; Ken Mason, Graeme Mason, and Abe Dorevich for instilling the love of dermatopathology during my dermatology residency, Claudia von Tscharner-who was the catalyst for my career path change, David Robson, and my dear friend Helen Power for their incisive editorial comments, and Ann Hargis, my co-author who was also one of my mentors and without whom this would have been a very different book.

I would also like to acknowledge the unconditional support of my husband, Ralf Mueller. Aside from being a truly wonderful partner, he is also an awesome critic.

Finally thank-you to Carroll Cann and Nicki Giandomenico at Teton NewMedia for encouraging and guiding me with this, my first book.

I will be allocating a portion of the profits of this book to research in the area of dermatopathology.

Sonya V. Bettenay

I acknowledge the family, friends, mentors, practitioners, and students who made the writing of this book possible, and the animals that gave me a reason to write.

Ann M. Hargis

Preface

The process of learning dermatopathology is long and often bewildering. This book is designed to facilitate that process for the practitioner in small animal medicine, residents in veterinary pathology and dermatology training programs, veterinary students, and veterinary pathologists in general practice. The book is also designed to be used as an aid to interpretation of skin biopsy reports. We have included tables and lists to enable busy practitioners to reference selected material quickly and efficiently.

The section on biopsy collection techniques is designed to improve knowledge of sample collection, handling, and submission for the ultimate purpose of establishing a specific diagnosis or differential diagnoses list for a skin disorder, and **to emphasize the critical role the clinical history plays** in histopathologic lesion interpretation and in establishing a specific diagnosis by the dermatopathologist. Two relatively new techniques to veterinary dermatopathology, the nail bed (claw bed) punch biopsy technique and the pinnal shave biopsy technique, are described.

The section on clinical lesion morphology is designed to help correlate clinical and histologic lesions and help practitioners select the biopsy technique most likely to provide a diagnostic biopsy sample for a variety of clinical lesions (*e.g.*, epidermal collarette, pustule, alopecia). The correlation of clinical and histologic lesions is extremely useful when formulating differential diagnoses lists and considering therapeutic options. It is a fundamental part of the learning of dermatology and dermatopathology.

The section on responses of the skin to injury is designed to explain the basic reaction of the different components of the skin to injury and to clarify or define terminology. This understanding is helpful when considering differential diagnoses, establishing a diagnosis and prognosis, and in the formulation of treatment regimes. The terminology used in dermatopathology is somewhat unique and the definitions will assist the general practitioner in the interpretation of biopsy reports and in communication with pathologists. The more experienced reader will recognize that we do not refer to the responses of the skin to injury as patterns. The reason to discuss responses of the skin to injury is that understanding the responses to injury is a first step in understanding patterns. For those interested in learning more about pattern terminology, we refer the reader to the following textbooks listed in the suggested readings section (Gross *et al*, Yager and Wilcock, Ackerman, Elder *et al*, and Hood *et al*).

The section on neoplasia is designed to provide a brief reference to help interpret biopsy reports and simply categorize neoplastic and non-neoplastic tumor-like lesions. Terminology and behavior of the basic categories of benign and malignant primary tumors of the skin are included. Tumor-like lesions are listed separately to differentiate them from tumors, to provide names and synonyms, and to provide basic definitions to aid in differential diagnoses of nodular lesions. The authors have largely based their terminology on the World Health Organization publications entitled, *"Histological Classification of Epithelial and Melanocytic Tumors of the Skin of Domestic Animals"* and *"Histological Classification of Mesenchymal Tumors of Skin and Soft Tissues of Domestic Animals.*

The section on laboratory techniques describes the basic laboratory procedures of sample processing, sectioning, and staining.

The glossary is designed to provide practitioners with easy access to definitions of terms often used in dermatopathology, and to illustrate a few of the potentially confusing lesions.

The section on differential diagnosis of clinical lesions is provided to help practitioners formulate clinical differential diagnoses based on clinical lesions and anatomic distribution patterns.

The appendices are designed to provide a biopsy submission form and sources for supplies useful in biopsy collection and handling. The case reviews illustrate how this book can be used to facilitate mananging the biopsy sampling process, formulation of clinical differential diagnoses, and pathology report interpretation.

Guide to Using This Book

Some Helpful Hints

The following icons are used in this book to indicate important concepts:

✓ **Routine.** This feature is routine, something you should know.

♥ **Important.** This concept strikes at the heart of the matter.

⌐ **Key.** This concept is a key one and is necessary for full understanding.

💣 Something serious will happen if you don't remember this.

✋ **Stop.** This doesn't look important but it can really make a difference.

Biopsy Technique

Readers wishing to increase the usefulness of the skin biopsy process will gain technical information pertaining to the when, why, where, and how of collecting skin biopsy samples. The surgical techniques for collecting the various types of biopsy samples (punch, incisional, excisional, claw bed, and pinnal) are described, as are methods of fixation and shipment of samples. For readers interested in learning about the techniques, a detailed text is provided. For a quick review prior to or during sample collection, simple checklists for each section are provided. This section also describes the importance of submitting a clinical history. To facilitate the process of submitting useful clinical history, an example submission form is provided in Appendix A.

Clinical Lesion Definitions and Biopsy Sampling Procedures

This section describes which biopsy technique is best suited for the various clinical lesions. It depicts primary and secondary clinical lesions, illustrates which biopsy technique is appropriate for which clinical lesion, and correlates the clinical lesion with examples of corresponding histopathologic lesions.

Response to Injury

Information on how the skin responds to injury, interpretation of terminology found in biopsy reports, and development of differential diagnoses for the various histopathologic lesions are described in this section. The section also correlates histopathologic lesions with clinical lesions to facilitate understanding of dermatopathology, and the inseparable nature of dermatology and dermatopathology.

Neoplasia

The location, terminology, and basic behavior of primary tumors of the skin are outlined in this section. This section also provides names, synonyms, and definitions of tumor-like lesions to aid in the differential diagnoses of nodular lesions of the skin.

Laboratory Techniques

Readers wishing to gain basic information pertaining to laboratory sample handling, processing, sectioning, and staining are referred to this section. A table of commonly used laboratory stains, including their uses, is listed.

Glossary and Appendices

The Glossary defines potentially unfamiliar terms that appear in the text or may appear in pathology reports. The Appendices provide information regarding suppliers of punch biopsy instruments, fixatives, tissue dyes, and tissue cassettes. The sample submission form is intended to be copied and used by practitioners. Case studies provide examples of how the information in this book can be used.

Differential Diagnosis of Clinical Lesions and Lesions in Selected Anatomic Locations

This section provides lists of clinical differential diagnoses to assist the practitioner in dermatologic evaluation, and to be used as a quick reference guide to finding lesions and disease descriptions in Sections 2, 3, and 4, and for selecting appropriate biopsy sampling procedures listed in Sections 1 and 2.

Section 1

Biopsy Collection:
Why,
When,
Where, and How

Biopsy Sample Selection

Why and When to Collect a Biopsy Sample

Collection of biopsy samples aids in the diagnosis of a skin disorder in several ways. Microscopic examination can lead to the diagnosis of a specific disease entity, can eliminate some of the clinical differential diagnoses, or may suggest a pathogenesis even if the specific disease entity is undetermined. Table 1-1 lists essential or important times to collect a biopsy sample.

Table 1-1
When and Why Biopsy Sampling is Recommended

WHEN	WHY
When skin lesions are acute and severe	To help identify a serious disease process so appropriate therapy can be instituted early
When therapy for a skin disorder is associated with significant side effects or may be life-threatening	To make certain the clinical diagnosis is accurate prior to initiating therapy
When a nodular lesion, ulcer, or chronic nonhealing lesion is present that could represent a tumor	To provide a histologic diagnosis and prognosis, so complete excision and other appropriate therapy can be instituted
When skin lesions appear unusual	To help identify a serious disease process so appropriate therapy can be instituted early
When the lesions are active, and prior to use of therapy that may alter the histologic character of the lesions	To provide superior diagnostic samples; lesions that have not been moderated by therapy
When lesions develop during the course of and while on therapy	To determine if there is an adverse reaction to drug therapy

Biopsy sampling may be useful

WHEN	WHY
When a skin disorder fails to respond to apparently appropriate therapy	To attempt to establish the cause of the skin disorder
When a skin disorder responds to therapy, but recurs when therapy is withdrawn	To attempt to establish underlying causes of the skin disorder
When there are multiple clinical differential diagnoses and the dermatological examination, including a thorough physical examination and evaluation of clinical pathologic analytes, cannot differentiate the conditions	To establish the diagnosis or eliminate some of the differential diagnoses quickly

Where to Collect a Biopsy Sample (Biopsy Site Selection)

Skin biopsy sampling is a useful diagnostic tool; however, too often results are "non conclusive." The lack of a specific diagnosis in part rests with sampling error, which can be prevented by adhering to basic principles of sample selection and collection.

The basic principles of sample selection and collection are:

✓ Spend at least 5 minutes examining the patient to select biopsy sites based on:
  - types of lesions present
  - the differential diagnosis list
  - knowledge of skin histology in the various anatomic sites of the body

Remembering that skin histology varies in different anatomic sites is important when selecting biopsy sites. For example, glabrous (non-haired) skin normally has fewer hair follicles and smaller sebaceous glands, so collecting skin from these areas in a suspected endocrinopathy makes histologic evaluation more difficult than using samples collected from skin that has more numerous hair follicles, such as the shoulder.

☍ Select multiple cutaneous sites representative of the range of lesions:

Most useful are fully developed non-treated primary lesions such as: macules, papules, pustules, nodules, neoplasms, vesicles, and wheals.

Also potentially useful are secondary lesions, such as: scales, crusts, ulcers, comedones, fissures, excoriations, lichenified areas, or scars.

Secondary lesions may be diagnostic or contribute substantially to the diagnosis. For instance, in pemphigus foliaceus, crusts are composed of old pustules, and the crusts usually contain acantholytic cells that facilitate diagnosis. Similarly, dermatophytes can be identified in crusts and lead to the diagnosis of dermatophytosis. Examination of the margin of a chronic ulcer may reveal a squamous cell carcinoma.

For diseases characterized by significant crusting, collecting crust **in addition to** representative biopsy samples of skin is recommended. Crusts are removed from the surface, placed within a piece of lens paper that is folded around the crust, and then placed in formalin fixative. The history form should include a note requesting that the crusts be processed for histologic examination ("please cut in crusts"). Crusts on the surface of biopsy samples can become dislodged from the surface and dissolve in the formalin prior to reaching the laboratory.

✓ Be aware that special types of lesions require special procedures.

Depigmentation: Select an area of active depigmentation (*e.g.*, gray) rather than the late stage (*e.g.*, white). Once the pigment has been lost from the epidermis, the process causing the pigment loss is over, and microscopic examination may only reveal that pigment is gone, but not the cause of the pigment loss. An incisional sample (ellipse or wedge) to include normally pigmented skin and the area of depigmenting skin is frequently useful as this technique provides the most active lesion (*e.g.*, the junction) for histopathologic evaluation.

Alopecia: Select samples from the center of the most alopecic areas. To avoid confusing the diagnosis, samples from junctional and normal areas should be placed in separate bottles and clearly labeled, or marked with colored ink.

Solitary nodule: Perform a wide excision as the biopsy technique of choice. Wide excision provides the entire specimen for histologic examination, allows margins to be examined should the lesion be a tumor, and can be curative in many cases. Surgical margins of 3 cm width are recommended for a suspected invasive neoplasm when possible.

Biopsy site selection is summarized in Table 1-2.

Table 1-2 Summary Checklist: Where to Collect a Biopsy Sample (Biopsy Site Selection)

✓	Spend at least 5 minutes looking for lesions based on: - Types of lesions present - Normal skin histology - Differential diagnoses
✓	Collect fully developed primary lesions
✓	Collect secondary lesions if they represent a significant portion of the disorder
✓	Collect additional crusts and wrap in lens paper, then place in formalin
✓	For pigment loss, collect gray areas and/or the margin of pigmented and nonpigmented skin
✓	For diagnosing alopecia, collect samples from the most alopecic areas and from partially hairless and normal areas for comparison purposes. Keep samples from most alopecic areas seprate from the others
✓	Label biopsy samples from the different areas using separate bottles or dyes, when appropriate
✓	For a solitary nodule, consider complete excision, which may effect a cure
✓	For suspected invasive tumors, 3 cm margins are indicated

How to Collect a Biopsy Sample (Biopsy Technique)
Biopsy Site Preparation

WARNING: The skin at a punch biopsy site should not be surgically prepared because surgical preparation may remove or damage the diagnostic portion of the sample. For an excisional biopsy of lesions deep to the epidermis, surgical preparation of the skin is acceptable.

Surgical Technique for Biopsy Collection

✓Gently clip or preferably cut hair with scissors, and gently remove hair.

✓For most skin biopsy samples, use manual restraint or sedation and local anesthesia. Inject lidocaine **without** epinephrine (1-2 mls) into the subcutis, ensuring that the entry point of the needle (26 gauge) is outside of the biopsy sample site to prevent artifactual changes to the tissue within the biopsy sample. However, injection of local anesthetic into the subcutis can**not** be used for suspected panniculitis lesions.

✓For biopsy procedures in some anatomic locations (*e.g.*, nasal planum, footpad, claw bed, eyelid), use general anesthesia to prevent patient discomfort or injury, and to obtain high quality samples not damaged during unexpected movement of the animal.

✓When collecting samples to evaluate for changes in follicular function, such as endocrinopathy, folliculitis, or follicular dysplasia, draw a line with a finely pointed permanent marking pen along the center of the biopsy sample site in the direction of the haircoat, (*e.g.*, from the head to the tail) (Figure 1-1). This line helps laboratory personnel section the sample to maximize histologic evaluation of hair follicles.

CLINICAL TIPS:

- Mark the biopsy site with four framing dots using a permanent marking pen before injection of the local anesthetic to be able to easily locate the anesthetized area.

- Allow 5 minutes for the local anesthetic to have an effect.

- For follicular disease, draw an orientation line, with permanent marking pen, along the center of the biopsy site in the direction of the haircoat.

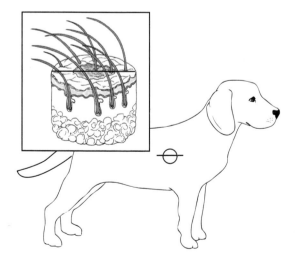

1-1 Diagram of biopsy sample site with line indicating direction sample should be cut in the laboratory (line extends from head to tail). This line essentially provides the laboratory technician with information on how to cut the section for proper orientation of hair follicles.

✓When a lesion in the panniculus is suspected, the punch biopsy technique may not extend deep enough into the lesion to collect an adequate sample. Therefore, if a deep lesion such as panniculitis is suspected, an incisional or excisional biopsy sampling procedure is recommended.

✓When collecting biopsy samples of suspected neoplastic lesions, there are two important considerations to help avoid implanting neoplastic cells into normal tissue. First, when multiple punch or incisional biopsy samples are collected from normal and suspected neoplastic sites, use different surgical instruments for normal versus neoplastic areas, or collect the samples from normal areas first and the neoplastic areas last. Second, if the incisional biopsy sampling procedure extends from lesional into adjacent normal tissue, and a neoplastic process is confirmed histopathologically, it is important that the subsequent surgical excision removes entirely the neoplastic mass as well as the biopsy-induced wound.

Punch biopsy sampling

Advantage: Quick and relatively atraumatic

Disadvantage: Causes rupture of large pustules or vesicles, makes sample orientation difficult or impossible if the punch sample includes both lesional and non lesional skin without drawing an orientation line, and is not suitable for lesions deep to the dermis, such as panniculitis.

Uses: For sampling small lesions that fit entirely within the diameter of the cutting surface of the punch instrument, or for diffuse lesions in which large areas of the lesional skin have a similar appearance. It is important to include 100% similar tissue within the punch instrument (*e.g.*, entire papule, diffuse area of erythema, diffuse area of lichenification (Figure 1-2).

Instruments required: Disposable punch biopsy instruments in 4, 6, and 8 mm diameter sizes (Appendix B). For sampling haired skin in most adult dogs and cats, 6 or 8 mm punch instruments are recommended; 4 mm punch instruments are most useful for biopsy sampling of footpads, nasal planum, or lesions near the eye or eyelids. Small punch biopsy instruments preferably should not be used to collect part lesional and part normal skin, because when the biopsy sample is cut in the laboratory for processing, the lesional area may not be included and lesional tissue may be damaged by the small punch instrument (Figure 1-3a). For certain lesions, and if precautions are taken, it is possible to collect part lesional and part normal skin using a large punch biopsy instrument. In this case, it is necessary to use an 8 mm punch instrument, and to mark a line on the sample with a fine tipped permanent marking pen in the direction the punch sample should be trimmed in the laboratory (Figure 1-3b). The line drawn on the sample instructs laboratory personnel how the sample should be trimmed to maximize evaluation of the junction between the lesion and normal skin. The line should be drawn only on samples in which the surface will not be damaged by the marking pen (*e.g.*, do not use with fragile surface scale) and in which the line will remain intact during fixation and transfer to the laboratory (*e.g.*, do not use on a friable, easily fragmented crust). Iris forceps (or small toothed forceps) may be used to gently grasp the non lesional base of the sample if necessary. Additional instruments that may be required include scissors, gauze sponges, and materials for wound closure.

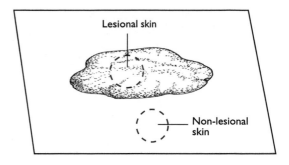

1-2 Diagram indicating method of collecting punch biopsy sample to include 100% lesional skin. This principle can be applied to obtain 100% lesional skin or 100% normal skin samples.

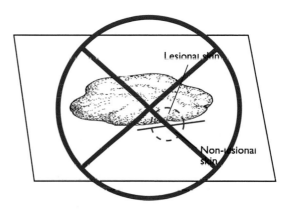

1-3a Diagram of punch biopsy sample collection in which half of the sample is normal skin and half is lesional skin. Note that, depending on way the sample is cut in the laboratory (a worse case scenario is depicted by the solid line), the normal portion of the skin may be the only portion evaluated microscopically. If only the normal half were included, the diagnosis could not be made.

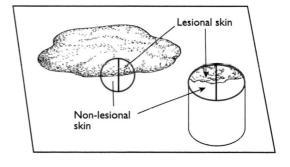

1-3b Diagram of punch biopsy sample collection in which half of the sample is normal skin and half is lesional skin. It is usually preferable, when using a punch instrument to select 100% lesional and 100% normal skin, but sometimes this is not possible. In that case, a line is drawn with a fine tipped marking pen to instruct the laboratory personnel the direction in which to trim the sample. The line should be drawn perpendicular to junction of lesion and normal skin so both will be included for the pathologist to review. Evaluation by the pathologist of the junction between normal and lesional skin may also be critical to identify the cause of the lesion.

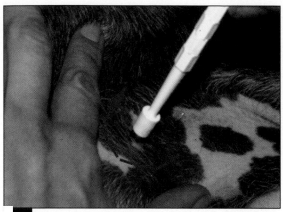

1-4 Photograph of collection of punch biopsy sample. The punch biopsy instrument is placed perpendicularly to the skin and is rotated in one direction only. The other hand is used to brace the skin.

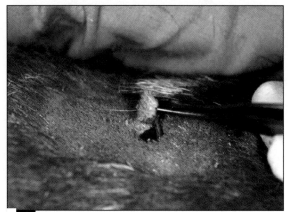

1-5 Photograph of collection of punch biopsy sample. When the lesion is in the epidermis or dermis, the sample may be grasped gently in a normal area at the base (subcutis) and a sharp scalpel or sharp scissors used to cut the sample from panniculus.

Technique:

- Hold the punch instrument with one hand at right angles to the surface of the skin and gently place over the selected lesion (Figure 1-4).

- Support the surrounding skin with the other hand.

- Apply firm continuous pressure and rotate the punch instrument in one direction only until the skin no longer "turns" with the rotation of the punch (there is an easing of tension while the punch instrument is turned).

- Gently grasp the base of the sample (the panniculus) with the forceps.

- Sever the subcutaneous attachments using a scalpel or scissors (Figure 1-5).

- Gently blot the blood from the biopsy sample with a gauze sponge, and place sample in 10% buffered formalin.

> **WARNING: Do not grasp the lesional area of the sample with a tissue forceps as this leads to "crush artifact" (Figures 1-6a and 1-6b).**
> **Crushed tissue may greatly hinder histologic evaluation.**

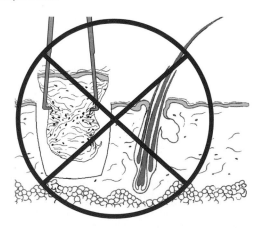

1-6a Diagram depicting the crushing of the dermis by the forceps. Subtle, or focal lesions may be completely destroyed. Depending on the extent of the crush artifact, there may be some residual tissue that can be evaluated, but histologic evaluation and interpretation are often significantly jeopardized.

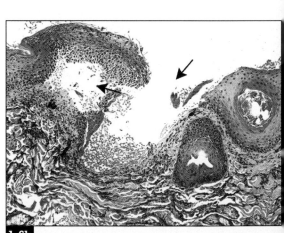

1-6b Photomicrograph of skin. The sample is not diagnostic due to crush artifact caused by grasping the epidermis and superficial dermis with tissue forceps with large teeth. The forceps has torn the epidermis, creating artifactual ulceration (arrow) and a subepidermal vesicle (arrow). Note that the arrows are pointing to "holes" in the tissue created by the teeth of the tissue forceps.

Incisional (Ellipse or Wedge) Sampling

Advantage: Technique allows evaluation of lesions where examination of the margin between normal and lesional skin is important, and when a larger sample of tissue is needed.

Disadvantage: More time consuming and technically difficult than punch biopsy procedure, and more likely to induce crush artifact (as forceps are used to grasp tissue) or artifact associated with curling or warping of sample in fixative.

Uses: Recommended when evaluation of the margin of the lesion is important, such as the margin of a depigmented lesion, the margin of an erosion or ulcer, or the margin of an epidermal collarette (Figure 1-7).

Instruments required: Scalpel, iris forceps (or small toothed forceps), and materials for hemostasis and wound closure.

Technique:

- Using a scalpel, make an elliptical incision with one narrow end located within lesional tissue and the opposite narrow end located within normal tissue.

- With the forceps, gently grasp a small portion of the normal part of the sample.

- Sever the subcutaneous attachments using a scalpel or small scissors.

- Carefully blot the blood from the biopsy sample.

- Place the sample, panniculus side down, onto a rigid piece of cardboard or broken tongue depressor (Figure 1-8a). This prevents the tissue from curling or warping (Figure 1-8b) when placed in the formalin, and optimizes the quality of the sample for interpretation by the pathologist.

- Place the biopsy sample and cardboard as a unit into 10% buffered formalin (tissue side down) (Figure 1-9).

- Close the wound routinely.

Excisional sampling

Advantage: Larger sample size facilitates collecting deeper lesions such as panniculitis, and solitary lesions that may be cured by complete removal.

Disadvantage: More time consuming and technically difficult than less invasive procedures. Usually requires general anesthesia.

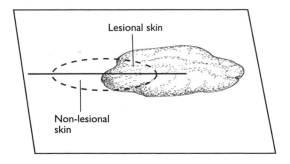

1-7 Diagram of method of collecting incisional (wedge, ellipse) sample to include lesion and margin of normal skin. This sample will be cut in the laboratory along the solid line. This method allows evaluation of the interface between normal and abnormal where a diagnostic lesion is most likely to be present. In alopecic conditions however, the most alopecic (and typically central) area frequently yields the most diagnostic sample.

1-8a Photograph of placement of thin incisional sample on cardboard to stabilize the sample while it fixes (hardens) to prevent the sample from curling in the fixative.

1-8b Photograph of a thin incisional sample that was not placed on cardboard and that curled in the fixative. It may not be possible for the technician to section this correctly.

1-9 Photograph of thin incisional sample adhered to cardboard being placed sample side down in fixative.

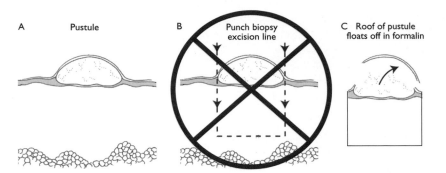

A   Pustule

B   Punch biopsy
excision line

C   Roof of pustule
floats off in formalin

1-10 Diagram of a large pustule collected with a small punch biopsy instrument. The roof of the pustule (and diagnostic part of sample) was lost because the biopsy punch instrument was too small.

Uses: For large pustules or vesicles that could be destroyed by punch or incisional sampling (Figure 1-10), for a lesion such as panniculitis that is deep to the dermis, or for a solitary lesion that could be cured by excision (Figure 1-11). If the lesion is deep to the epidermis, surgical preparation is acceptable. If a central lesion involving the epidermis is present, surgical preparation around the margin of the lesion, leaving the lesion intact, may also be possible. Use care not to damage the centrally located lesion.

Instruments required: Scalpel, iris forceps (or small toothed forceps), scissors, hemostats, and others as dictated by extent of surgery and wound closure.

Technique:

- Make an incision through clinically normal tissue using a scalpel.

- Use 3 cm margins around the nodular mass if an invasive neoplasm is suspected. Depending on the depth of the mass, it may be necessary to incise through the panniculus and underlying musculature.

- Gently grasp the specimen with the forceps in non lesional skin and remove the specimen. Close routinely.

- For a large nodule, section through the lesion to allow formalin to penetrate into the tissue for proper fixation. If a large nodular mass is not incised prior to being immersed in fixative, much of the central portion of the mass will autolyze. Only a narrow zone of tissue along the circumference of the mass will be adequately fixed, and microscopic examination will be significantly hindered.

- Nodules that are relatively small (less than 3 cm) can be left intact.

Moderately sized nodules (between 3 cm and 6 cm), may be incised once through the epidermal surface, but leave the deep pannicular margin intact (Figure 1-12). Formalin can then penetrate into the incised surface of the nodule, but the deep margin remains intact to enable examination of this margin for completeness of removal. For large masses (greater than 6 cm), make multiple incisions through the mass (leave the deep pannicular tissue intact if possible) to facilitate fixation and maintain orientation of the nodule within the sample.

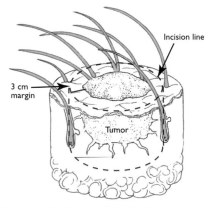

Incision line

3 cm
margin

Tumor

1-11 Diagram of collection of nodule by excisional sampling. If possible, 3 cm margins should be collected especially if the nodule is a suspected neoplasm.

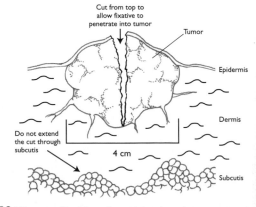

Cut from top to
allow fixative to
penetrate into tumor

Tumor

Epidermis

Dermis

Do not extend
the cut through
subcutis

4 cm

Subcutis

1-12 Diagram of incision of a nodular dermal mass measuring between 3 and 6 cm. Incision of the nodule through the epidermal surface leaving the subcutaneous surface intact allows the formalin fixative to penetrate the nodule for proper fixation while leaving the subcutaneous margin intact so the deep margins can be evaluated for completeness of removal.

WARNING: Incise a large nodule in 2 cm slices perpendicular to the surface starting at the epidermis. Leave about 1 cm of the deep margin intact and not incised. If the nodule is incised from the bottom, the fat and muscle will retract away from the deep margin of the mass and the mass will bulge toward the incision line. Incising a nodule from the bottom makes it nearly impossible for the pathologist to examine the mass for completeness of removal.

Canine Claw Bed Biopsy Procedure Developed by Mueller and Olivry

Advantage: Provides a method to obtain claw bed biopsy samples without digital amputation. Surgical time is short, (just 5 minutes, when personnel are experienced with the procedure).

Disadvantages: Requires general anesthesia and good intra-operative and post-operative analgesia. Significant hemorrhage can occur. Laboratory personnel need to be provided with information pertaining to the orientation of the sample; consider inking the haired skin margin. Does not provide as much tissue as digital amputation so sample may be inferior to that obtained by digital amputation.

Use: For diseases of the canine claw bed and formation of abnormal claws. Affected dew claws, if present, are preferred to weight-bearing digits as dew claws are less likely to be associated with post biopsy lameness. Ideal samples are those with affected claw remaining, and no evidence of secondary suppurative inflammation.

Instruments required: Usually 8 mm punch biopsy instrument unless the patient is small, scalpel, forceps, tourniquet, materials for control of hemostasis and wound closure. A new (previously unused) punch instrument **is essential**.

Technique:

- Surgical preparation is **not** recommended. Use of a tourniquet will reduce hemorrhage during the procedure.

- Position the punch biopsy instrument at the lateral edge of the claw, so the cut is parallel with the claw, and includes a thin "shave" sample of the horn of the claw, claw bed skin, and digital bone (Figure 1-13). It may be helpful to first cut a small notch in the claw to serve as a guide and to help keep the punch instrument from slipping off the claw.

- Rotate the biopsy instrument in one direction only cutting deep into the tissue (moderate physical force is required to move the biopsy instrument through the claw and digital bone). Note that it is not feasible to use this technique on healthy claws. The punch instrument will likely break.

- Cut the base of the sample with a scalpel and gently blot blood from the sample. Mark the skin surface with dye to orient the sample in the laboratory. Indicate on the submission form how the sample was marked. Gently transfer the sample to 10% buffered formalin fixative.

- Use two deep sutures in the skin to close the wound. Digital pressure facilitates hemostasis.

CAUTION: **Use digital amputation if the soft tissue or bone swelling is significant, indicating that osteomyelitis or a tumor may be present (*e.g.*, when the lesion represents more than a claw growth abnormality).**

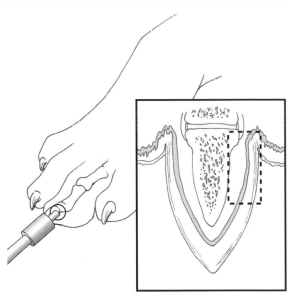

1-13 Diagram of Mueller and Olivry's biopsy procedure for abnormalities of the claw or diseases causing loss of claws. The punch biopsy instrument collects a sample from either the lateral or medial edge of the claw including the claw bed, skin and portion of third phalanx.

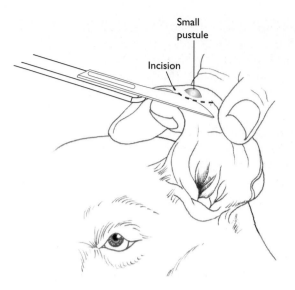

Small
pustule

Incision

1-14 Diagram of the shave biopsy technique using the pinna as an example. The scalpel blade cuts through the superficial dermis. This type of sampling provides small samples of superficial (e.g. epidermal) lesions, but is not appropriate for lesions of adnexa, dermal vasculature, or deeper dermal tissue.

The Shave Biopsy Technique

Advantage: Provides a quick method to collect multiple samples of superficial lesions without inducing cosmetic damage to the animal. (Figure 1-14). Particularly useful for the inner (concave) surface of the pinnae and for the nasal planum of cats with suspected early squamous cell carcinoma.

Disadvantage: Does not allow collection of larger epidermal lesions such as large pustules, or deeper lesions in the dermis, such as vasculitis.

Uses: For collecting multiple samples of small superficial papular and pustular lesions of the pinnae (or other similar surfaces) quickly and without inducing cosmetic damage.

Technique (for pinna):

- Use general anesthesia, or sedation with local anesthesia.

- Fold the pinna (with the lesion on the surface of the fold).

- Use a scalpel to slice (shave) a portion of lesional epidermis and superficial dermis from the top of the fold (Figure 1-14). The deep dermis and cartilage remain untouched.

- Use digital pressure to stop bleeding. Suturing is usually not possible.

- Gently place the small samples within a piece of lens paper, and then fold the lens paper and place in formalin fixative. Enclosing the samples within lens paper keeps the samples from being lost or overlooked by the laboratory.

Although the samples are small, the fact that multiple areas can be easily and quickly sampled, makes this a useful technique for collecting small superficial pinnal lesions for histopathologic evaluation. For large epidermal lesions (large pustules) or lesions that are deeper in the ear tissue (*e.g.*, vascular, dermal, adnexal, or cartilage), punch or incisional biopsy sampling is required.

Identifying or Marking Biopsy Samples and Surgical Margins

✓When collecting multiple samples, individual samples can be identified by placing them in separate labeled bottles, by placing different numbers or colors of sutures in a non lesional area (usually subcutis), or by using differently colored dyes made for marking tissues (see Section 5 and Appendix B). Small punch biopsy samples or thin incisional (ellipse or wedge) samples also can be placed in labeled tissue cassettes (see Section 5 and Appendix B) and then into fixative. The history form should clearly document the identity of samples so marked.

✓Marking the surgical margin of an excisional biopsy sample with dye specifically designed for tissue can help the pathologist evaluate the margin for completeness of removal. Care should be taken to ensure that the entire surgical margin is evenly covered with the dye.

✓Tissue dyes also can be used to help orient claw bed punch biopsy samples for processing in the laboratory. Typically, the skin side (non-cut outside surface) of the sample is marked with dye. It is important to specify on the history form exactly where on the sample the dye was applied. Alternatively, the orientation of the sample can be facilitated by placing the sample into a tissue cassette (see Section 5) with the cut surface facing downward, and then placing the cassette into fixative. Again, the history form should contain information on exactly how the sample was placed in the cassette (e.g., cut surface facing down).

☞ When using differently colored dyes for marking tissues, gently pat the sample dry to remove blood and fluid, apply the dye with a cotton-tipped applicator (Figure 1-15), and wait until the dye no longer is shiny, (about 10 minutes), and transfer the sample to fixative. If the sample is placed into the fixative while the dye is wet, the dye may leach through the fixative onto the other samples in the bottle.

Techniques for collection of biopsy samples are summarized in Table 1-3.

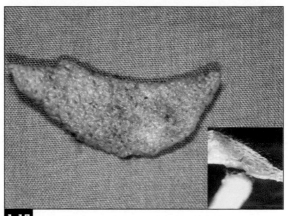

1-15 Photograph of skin sample with yellow dye painted over the surface. Once the ink is dry, the sample can be placed in the bottle of formalin with the unmarked samples. Inset: tip of cotton swab is used to apply the dye to tissue.

Table 1-3 Summary Checklist: How to Collect a Biopsy Sample (Biopsy Technique)

✓	Gently clip or cut hair with scissors, and carefully remove hair
✓	Mark site of biopsy sampling with framing dots using permanent marking pen; for follicular diseases, draw an orientation line with a fine tipped permanent marking pen along the center of the biopsy sample site in the direction of the hair coat
✓	Inject local anesthetic into subcutis and wait 5 minutes. Use general anesthesia for claw bed, footpad, nasal, pinnal, or eyelid samples
✓	For punch biopsy sampling: - Select small lesions that fit entirely within the cutting surface - Select diffuse lesions in which lesional skin has similar appearance - Use 4, 6, or 8 mm biopsy punch appropriate for sample and site - Rotate punch biopsy instrument in one direction only - Gently grasp sample at base with iris forceps - Excise attached sample from skin with scalpel or small scissors - Gently blot blood from sample and place in fixative
✓	For lesions in which examination of the junction of affected and normal skin is important: - Ideally use incisional (ellipse or wedge) biopsy sampling - Alternatively, use an 8 mm punch biopsy instrument and draw a line on the sample with a fine tipped permanent marking pen in the direction the sample should be trimmed in the laboratory to facilitate evaluation of the junction between normal and abnormal (see Figure 1-3b)
✓	Use excisional sampling for: - Lesions such as large pustules or vesicles that could be damaged by other sampling techniques - Lesions deep to the dermis such as panniculitis - Nodular lesions that could be cured by complete excision
✓	To avoid implanting neoplastic cells into normal tissue during sampling procedures of suspected neoplastic lesions: - Biopsy normal areas first or use different surgical instruments for sampling normal versus suspected neoplastic tissue - After the initial incisional biopsy procedure at the margin of a neoplasm, make certain the subsequent surgical excision removes the entire neoplastic mass as well as the biopsy induced wound
✓	For nodular lesions greater than 6 cm in diameter, incise at 2 cm increments through the epidermal surface leaving subcutis and musculature intact. This facilitates fixation, but allows the deep margin to be examined for completeness of removal
✓	For thin samples prone to curling in the fixative: - Gently blot blood from sample - Place sample on a firm substance such as stiff cardboard or a piece of tongue depressor - Let sample adhere to the firm substance for about 30 seconds - Immerse tissue-side down into fixative
✓	For diseases characterized by crusting, include additional crusts with biopsy samples. Request processing of crusts on the submission form
✓	For claw bed biopsy samples, select affected dew claws or affected digits - Use general anesthesia with good intra- and post-operative analgesia - A tourniquet reduces hemorrhage during the procedure - Rotate, in one direction only, a punch biopsy instrument so as to include a thin piece of lateral claw, claw bed skin, and digital bone - Cut base of sample with a scalpel

continued

Table 1-3 **Continued**

✓	For superficial papular and pustular lesions of the pinnae
	- Gently fold the pinna with the lesion on the top of the fold
	- With a scalpel, slice (shave) a portion of the lesional epidermis and superficial dermis from the top of the fold
	- Hemostasis is accomplished by digital pressure
	- No suture is required
	- Place small samples within lens paper, gently fold the lens paper, and place in formalin fixative
✓	To identify multiple samples:
	- Place samples in separate bottles
	- Place different numbers or colors of sutures in nonlesional areas
	- Use dyes made for marking tissues (see Section 5)
	- Place small samples, crusts, and claw bed punch biopsy samples in labeled tissue cassettes (see Section 5)
✓	To identify sample margins
	- Mark the surgical margin with dye
	- Dyes also help orient claw bed punch biopsy samples

Biopsy Sample Fixation (How To Preserve a Biopsy Sample)

Standard Fixation Techniques

✓ Punch biopsy specimens should be placed in 10% neutral buffered formalin (NBF), the standard fixative for histopathology. The volume of formalin should be 10 times the volume of the tissue samples. To avoid warping in the fixative, thin incisional and excisional biopsy specimens should be gently attached to a flat object, such as a piece of tongue depressor, and allowed to dry for about 30 seconds to adhere to the object, then immersed in formalin (see Figures 1-8 and 1-9). In cold climates during winter months, adding 1 part alcohol to 9 parts 10% NBF reduces chances of freezing of the specimens during transport. Freezing can disrupt tissue (Figure 1-16a and 1-16b).

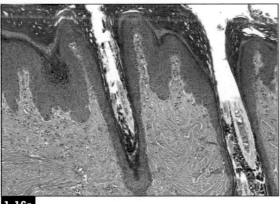

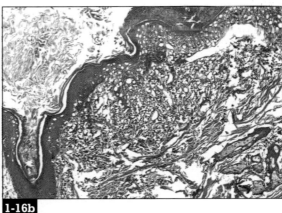

1-16a   1-16b

Photomicrographs of skin. In figure 'b' the sample froze during transit. Figure 'a' is taken from non-frozen tissue. Ice crystal formation reduces the ability to evaluate the tissue microscopically.

Fixation for Suspected Immune Mediated Skin Diseases

✓For suspected immune mediated skin diseases such as pemphigus, lupus erythematosus, and bullous pemphigoid, direct immunostaining procedures that evaluate for immunoglobulin or complement deposition in the skin are available. Unfortunately, these immunostaining techniques may provide false positive or false negative results, minimizing their diagnostic value and reducing the frequency of their use by veterinary dermatologists and dermatopathologists. Because of the potential of false negative and false positive results, **immunostaining, if performed at all, always should be evaluated in conjunction with conventional histopathology, never as a solitary procedure**. Newer techniques that allow for more specific antigen targeting such as a desmosomal protein (desmoglein), use of "salt-split" skin sections for subepi-dermal bullous diseases, use of better substrates for indirect immunostaining, and use of immunoblotting and ELISA tests are being developed to facilitate diagnosis of immune mediated skin diseases. When widely available at laboratories, these techniques should improve diagnostic accuracy for immune mediated skin disease in the future.

✓Currently, if immunofluorescence evaluation is desired, specimens should be placed in Michel's medium, which preserves immunoglobulin and complement. Immunohistochemistry (immunoperoxidase) staining is performed on formalin fixed samples; however, for best results, samples should not remain in formalin longer than 48 hours. Prolonged fixation in formalin results in cross-linking of protein and false negative results. Section 5 contains more detail on these special staining techniques.

Fixation for Selected Tumor Studies or for Selected Infectious Agents

✓Special studies for the identification of cellular antigens are also available. These studies are usually performed after initial cytologic or histopathologic evaluation, which suggests special studies may help to reach a more specific diagnosis. These studies are particularly useful in the diagnosis of tumors or tumor-like lesions (see Section 5). Depending on the antigen, either fresh tissue or formalin fixed tissues can be used, but fresh tissue is usually preferred, especially when evaluating for the cluster of differentiation (CD) antigens on the surface of leukocytes and other cells. Contact the laboratory for shipping and handling recommen-dations, and alert the laboratory when the samples are shipped. To ship fresh tissue samples, the samples should be wrapped in saline soaked gauze, placed in a sturdy plastic bag with frozen cold packs, and delivered to the laboratory within 24 hours. Delivery by local courier or commercial express shipment is recommended.

✓Diagnosis of infectious agents can be accomplished by histochemical stains on formalin fixed tissues, or for selected infectious agents, immunostaining procedures (see Section 5). Contact laboratory personnel, regarding the preferred method of fixation for the immunostaining procedure for the infectious agent in question. Tips for fixation of biopsy samples are summarized in Table 1-4.

Table 1-4 Summary Checklist: Sample Fixation (How to Preserve a Biopsy Sample)

✓	Use 10 times the volume of fixative for each volume of sample
✓	For routine histopathology use 10% neutral buffered formalin (NBF)
✓	Place thin samples (prone to curling or warping) onto a piece of cardboard or tongue depressor to keep the sample straight and flat, then immerse the sample into the fixative
✓	For nodular lesions greater than 6 cm in diameter, incise at about 2 cm increments through the epidermal surface leaving subcutis and musculature intact. This facilitates fixation, but allows the deep margin to be examined for completeness of removal
✓	To help prevent freezing in transit during cold weather: - Add 1 part of alcohol to 9 parts of 10% neutral buffered formalin - Use an insulated shipping container - Ship the specimen by an express mail or courier service
✓	For immunofluorescence (immune mediated skin disease) - Use as an additional diagnostic test only (does not replace standard histopathology) - Place one or two samples in Michel's media
✓	For immunohistochemistry (immune mediated skin disease or cellular antigens associated with tumors, tumor-like conditions, or infectious agents) - Place samples in 10% formalin and ship samples so they are in formalin less than 48 hours - Some evaluations for cellular antigens or infectious agents require submission of fresh tissue. Contact laboratory personnel for shipment instructions. - When in doubt about the proper method of fixation or shipment, consult with the laboratory personnel performing the evaluation

Clinical History (Why Provide a Clinical History)

✍ The histopathologic diagnosis and lesion interpretation by a pathologist usually require knowledge of the clinical features of the lesions. In dermatopathology, perhaps more than any other field, a large percentage of the diagnosis rests with the history and signalment. **Therefore, it is essential to include a thorough clinical history with the biopsy samples** (see Appendix A for sample submission form). The minimum clinical information that must be provided is: age, breed, sex of the animal; location of lesions and samples, gross appearance, duration of lesions; and the presence or absence of symmetry and pruritus. Results of clinical laboratory evaluations, current medications (*e.g.*, steroids may decrease eosinophil number), and response or lack thereof to therapy should be included along with a list of clinical differential diagnoses.

✍ The importance of providing the clinical history cannot be overemphasized. For instance, presence of luminal and mural folliculitis with no apparent follicular infectious agents in H&E-stained sections in conjunction with the history of lack of response to appropriate antibiotic therapy would prompt the pathologist to request a fungal stain to help identify a dermatophyte infection. Without the history of lack of response to appropriate antibiotic therapy, the folliculitis could easily be presumed to be of bacterial origin, and the fungal infection missed.

Sample Shipment (How to Ship Samples to the Laboratory)

✓ Many veterinary diagnostic laboratories provide courier service, simplifying biopsy sample delivery. If samples are shipped commercially, precautions must be taken. Formalin is a hazardous substance, so leakage must be prevented. The leaked formalin can render other samples (*e.g.*, for microbiology) useless, may result in biopsy samples becoming dehydrated, or may result in samples being detained or confiscated by the shipping service. Use of crush proof and leak proof containers is requisite. It is helpful to secure bottles containing formalin or other hazardous materials within additional, non-leakable, containers, and to use sturdy, crush proof outer containers.

✐ If shipping during especially cold weather, using insulated boxes and shipping by express services can help prevent freezing. Freezing of a sample while it is immersed in liquid (formalin) results in formation of large ice crystals in the tissue, usually rendering the sample non diagnostic (see Figure 1-16).

✓ If a patient has a serious disease, it is advisable to ship the sample by an express service to avoid shipment-related delays. If a patient is seriously ill and it is believed the biopsy sample holds information critical to the management of the case, alert the diagnostic laboratory and pathologist to insure they are able to expediently process and evaluate the sample.

✓ Addresses and personnel change, so if your veterinary hospital uses a specialty laboratory infrequently, check prior to sample shipment to make certain that the address is correct and service is available. As an added safeguard, always use a return address, and consider shipping the sample by a service that provides a tracking number so the sample can be traced if necessary. Tips for shipping biopsy samples are summarized in Table 1-5.

Table 1-5 Summary Checklist: Sample Shipping (How to Ship a Biopsy Sample)

✔	Place samples in leak proof, sturdy, unbreakable container, and tighten the lid. When samples are shipped via air, the increased pressure can force formalin out of the container
✔	If concurrently shipping samples for other procedures, such as microbiology, double wrap the formalin samples in nonleakable, nonbreakable material to prevent formalin from contacting the other samples
✔	Use sturdy mailing tube or box that is crush proof
✔	If you use a specialty diagnostic service infrequently, check the shipping address, use a return address, and if a specific pathologist's evaluation is sought, check their availability prior to shipping
✔	Use insulated box if shipping when samples might freeze
✔	If patient is seriously ill, ship the sample by an express service
✔	Consider shipping with a service that provides a tracking number

Microscopic Interpretation of a Biopsy Sample

✐ The skin has a limited number of ways to react to the variety of insults to which it is exposed (infections, traumatic events, chemical irritants, allergens); therefore, many skin disorders share the same microscopic changes. Classic or "text book" histopathologic lesions are seen occasionally, but as in the clinical situation, classic features are not always present in biopsy samples at one point in time. Careful completion of an appropriate skin biopsy history form greatly improves the chances the pathologist interprets the histologic lesions accurately, and helps establish a specific diagnosis. Many dermatoses are diagnosed using a combination of the signalment (age, breed, sex), clinical presentation (lesion distribution, morphology, duration), response to therapy, and supportive histopathology (Table 1-6). Therefore it is essential to include with the biopsy sample a thorough clinical history (see Appendix A). The histopathologic examination alone provides a morphologic diagnosis (superficial pustular dermatitis with a few acantholytic cells). The history helps determine the significance of this morphologic diagnosis, can determine if the samples are representative of the clinical lesions, and if rebiopsy or special stains are necessary, and if there has been a "sample mix up".

✓ Special staining procedures, typically performed on histologic sections after evaluation with H&E staining, are used by pathologists to enhance differential staining of cellular constituents such as mast cell granules, interstitial constituents such as elastic fibers, or etiologic agents such as bacteria or fungi. These special staining procedures require additional time and may delay the final pathology report. Commonly used special stains are listed in Section 5.

Table 1-6 Example of Importance of History in Interpretation of Microscopic Lesions

CLINICAL HISTORY*	HISTOPATHOLOGIC FINDINGS**	INTERPRETATION (principal differential diagnosis)
Terrier cross, **5-month old**, female with nonpruritic, **nonfollicular pustules** and a few crusts in the **axillary and inguinal areas.**	Superficial pustules with no visible bacteria and an occasional acantholytic cell. Pustular crusts also present.	Staphylococcal pyoderma (impetigo)
Chow chow, 4-year old, female has **thick crusts and pustular lesions** on the **face, nasal planum, pinnae,** and disseminated in other sites; there is **footpad crusting**; the dog is febrile; there has been **no response to 3 weeks of cephalexin.**	Superficial pustules with no visible bacteria and a variable number of acantholytic cells. Pustules are large and bridge follicles. Pustular crusts are thick and multilaminated.	Pemphigus foliaceus
Chow chow, 4-year old, female has **thick crusts and pustular lesions** on the **face, nasal planum, pinnae,** and disseminated in other sites; there is **footpad crusting**; the dog is febrile; there has been **no response to 3 weeks of cephalexin.**	One sample submitted. Superficial pustule containing cocci, but no acantholytic cells.	Dog has pyoderma, but the **major diagnosis was missed** because only one sample was submitted, and the disease covering the rest of the dog was not sampled. The diagnosis of the principal disease cannot be made without rebiopsy.
Chow chow, 4-year old, female has **thick crusts and pustular lesions** on the **face, nasal planum, pinnae,** and disseminated in other sites; there is **footpad crusting**; the dog is febrile; there has been only mild **partial response to 3 weeks of cephalexin, discontinued 3 weeks ago.**	Superficial pustules with coccoid bacteria in pustules and a variable number of acantholytic cells. Some pustules are large, bridge follicles, and contain more numerous acantholytic cells. Pustular crusts are thick and multilaminated. Acantholytic cells and cocci also present in crusts.	Pemphigus foliaceus complicated by staphylococcal pyoderma.
Adult male mixed breed dog had a **few pustules and epidermal collarettes on the abdomen that responded in the past to trimethoprim potentiated sulfa.** This time the pustules **initially responded to this therapy, but then suddenly worsened and spread over the entire body.**	Superficial pustules with no visible bacteria and a variable number of acantholytic cells, some in clusters (rafts). Few pustular crusts.	Pemphigus foliaceus, probably drug associated, and complicating the antecedent staphylococcal pyoderma.
Dalmatian, 6-year old male, neutered with symmetrical **crusting, alopecia, and pustules** on the haired skin of the face, **sparing the nasal planum**, and **asymmetrical crusts and alopecia on the dorsal back.** **Lesions have not responded to enrofloxacin or cephalexin.**	Superficial pustules with no visible bacteria and a variable number of acantholytic cells. Pustular crusts also present. Fungal stain revealed spores in and around the hair shafts.	Dermatophytosis
Pomeranian, 5-year old intact **male with noninflammatory, symmetrical alopecia** and hyperpigmentation on the caudal trunk, and **nodule on left shoulder.**	Superficial pustules with no visible bacteria and a variable number of acantholytic cells. Few pustular crusts.	Sample mix up

* Items in boldface are important clinical clues that assist interpretation of histologic lesions.
**Histopathologic findings: note how similar the histologic lesions are in each case. The history helps the pathologist determine if the samples are adequate and are appropriate for the animal, if special stains are necessary, and facilitates interpretation of the lesions.

Limitations of Dermatopathology (What a Dermatopathologic Evaluation Cannot Do)

✓It cannot replace a thorough clinical evaluation of the skin, but can substantially contribute to that evaluation.

❤ It cannot substitute dermatopathology expertise for insufficient history or insufficient or unrepresentative biopsy samples.

☞ It may not be able to identify the specific disease present. However, the evaluation can provide a list of differential diagnoses and/or rule out diagnoses from the clinical diagnostic differential list. For instance, histopathologic evaluation can often identify the source of the problem as hypersensitivity, but cannot identify the specific hypersensitivity (food versus contact). Similarly, histopathologic evaluation can identify a pustular disease process, but may not be able to differentiate between the causes of pustular disease, (especially without history, adequate biopsy samples, and lesions at the appropriate stage of disease).

Value of Referral to or Consultation with a Dermatology Specialist

✓Veterinary dermatology is a specialty in which residents spend a minimum of two years seeing dermatology cases exclusively, collecting and evaluating biopsy samples with guidance of a dermatopathologist, studying the dermatology literature, and preparing for board certification. Most veterinary dermatologists limit their professional careers to the practice of veterinary dermatology. Many benefits for the practitioners and their patients result from professional interaction with a veterinary dermatologist. These benefits include obtaining an accurate diagnosis, an appropriate therapeutic protocol, and ongoing patient management guidelines.

✓In regard to veterinary dermatopathology, specialty boards are not established. However, an association called, the International Society of Veterinary Dermatopathology (ISVD), has been formed, and is composed of veterinary dermatologists and veterinary dermatopathologists who have a special interest and expertise in skin disease. The objectives of the ISVD include the advancement of veterinary and comparative pathology through education, adapting and implementing emerging techniques for morphologic diagnosis, and communication and interaction of individuals who have a professional interest in histologic interpretation of skin disease in animals. Membership in this association illustrates an interest in, and often a commitment to, the advancement of the field of veterinary dermatology. Submitting skin biopsy samples to a veterinary pathologist with special interest in skin disease (dermatopathologist) is appropriate; however, the expertise of the dermatopathologist cannot substitute for the lack of a clinical history or for poor quality biopsy samples. The most knowledgeable dermatopathologist may not be able to determine the diagnosis if an inadequate sample or inadequate history are submitted.

Ancillary Procedures to Facilitate Diagnosis

When an animal is sedated for biopsy sample collection, it is an ideal time to collect samples for other diagnostic procedures.

Examples include:

✓Aspiration of pustules (for pustular cytology)

✓Performing touch imprints of the cut surface of a suspected neoplastic or infectious lesion (for submission for cytologic evaluation).

✓Surgically preparing a site and, using aseptic technique, collecting a tissue sample of a nodular lesion that may have an infectious etiology (for microbiology submission)

✓Collecting additional samples for cellular antigens (to aid in the diagnosis of tumors or tumor-like conditions or for aiding in the diagnosis of immune mediated diseases)

Common Errors of Sample Collection and Submission ("Do's" and "Don'ts" of Biopsy Sample Collection and Submission)

For convenience, Table 1-7 summarizes the common errors associated with sample collection and submission. Tables 1-8 and 1-9 summarize the do's and don'ts of sample collection and submission.

Table 1-7
Common Errors in Sample Collection or Submission

Collecting only one sample or collecting samples that are too small (<4 mm)
Surgically preparing or otherwise damaging the skin surface prior to or during biopsy sampling
Not providing a dermatological history
Waiting until lesions are too chronic before collecting biopsy samples
Treating the animal with anti-inflammatory therapy (*e.g.*, glucocorticoids) prior to collecting samples
Using the wrong biopsy technique for the lesion (*e.g.*, small punch instrument for large pustule)
Crushing sample with tissue forceps
Coagulating small sample with cautery/laser
Failing to collect surface crust
Improperly fixing the sample, such as not using cardboard to provide structure for a thin incisional or thin excisional sample, using the wrong fixative, not using enough fixative, letting a sample in formalin freeze

Table 1-8
Do's of Sample Collection and Submission

Include a thorough dermatological history with a list of differential diagnoses
Biopsy early
Collect multiple samples representative of the range of lesions present and include crusts
Biopsy before using anti-inflammatory therapy, or wait for appropriate "withdrawal" period prior to biopsy
Use correct biopsy procedure for the type of lesion
Label samples from different areas using separate jars if necessary
Handle samples gently and fix properly
Ship to avoid leaking fixative, crushing or freezing of sample, and to prevent delays

Table 1-9
Don'ts of Sample Collection and Submission

Do not scrub or otherwise clean the skin surface
Do not use tissue forceps to grasp the lesional area of the specimen
Do not use a biopsy instrument that is too small (4 mm diameter is minimum useful diameter)
Do not squeeze specimen
Do not use laser/cautery techniques on small samples or if it is important to examine margins for tumor cells
Do not use alternative fixatives; use only 10% buffered formalin for standard histologic evaluation (no pure alcohol, water, freezing)
Do not expect microscopic examination of a few hastily collected samples to replace a thorough dermatologic evaluation

Section 2

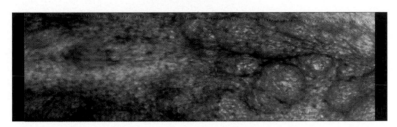

Clinical Lesion Definitions,
Recommended Methods of Biopsy Sampling,
and Representative Histologic Lesions

Clinical Lesion Definitions, Recommended Methods of Biopsy Sampling, and Representative Histologic Lesions

The information in this section is assembled in a tabular form to allow easier visualization of the clinical and comparable histologic lesions. The first column illustrates dermatologic lesions and provides a brief definition of the lesion. The second column illustrates the same dermatologic lesion and the authors' recommended biopsy-sampling technique. The sampling techniques were chosen based on the unique features of lesions, artifacts likely to be introduced by sampling method, simplicity of biopsy technique, and time efficiency. For many dermatologic lesions, incisional (wedge or elliptical) biopsy samples are as efficacious as punch biopsy samples; however, only punch samples have been recommended or illustrated. This is because use of disposable punch biopsy instruments markedly facilitates collection of high quality biopsy samples with minimal surgeon-induced artifacts in a quick efficient manner, whereas incisional biopsy samples are more labor intensive and can be associated with surgically induced artifacts. However, the biopsy method of choice is often one of personal preference. For some dermatologic lesions, the incisional (elliptical wedge) biopsy sampling procedure is **preferred** over the punch biopsy method, particularly where histologic evaluation of the transition from normal to abnormal skin is critical for diagnosis. In those instances, the incisional method is illustrated. Procedures for collecting punch, incisional (elliptical or wedge), and excisional biopsy samples are described in Section 1. The third column contains sketches of histologic lesions that exemplify the corresponding dermatologic lesions, and provides names of diseases in which the histologic lesion may be a feature. This list of diseases is not intended to be complete. The fourth column contains photomicrographs of examples of the corresponding dermatologic lesions, stained with hematoxylin and eosin.

Table 2-1
Clinical Lesion Definitions, Recommended Methods of Biopsy

CLINICAL LESION DEFINITION	RECOMMENDED METHOD OF BIOPSY SAMPLING IS ILLUSTRATED

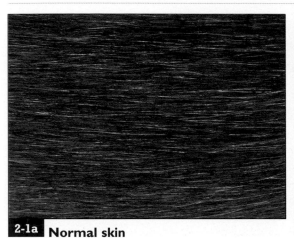

2-1a **Normal skin**

2-1b Punch biopsy samples; however, a variety of techniques are acceptable

A line is drawn through sample to illustrate to laboratory personnel how to cut the sample

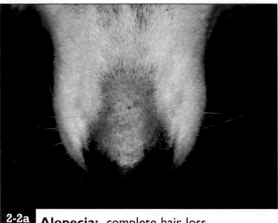

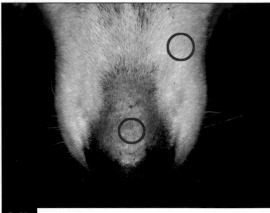

2-2a **Alopecia:** complete hair loss

2-2b Punch biopsy samples of completely alopecic areas

Punch biopsy samples of normal areas labeled and kept separate from the completely alopecic areas

Punch biopsy samples; however, variety of techniques acceptable

Sampling, and Representative Histologic Lesions

SKETCH OF LESION AND COMMON EXAMPLES

PHOTOMICROGRAPH OF EXAMPLE LESION

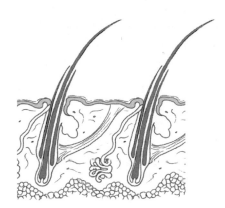

2-1c

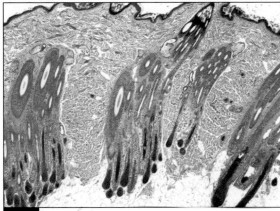

2-1d Note active hair follicles and hair shafts in follicles.

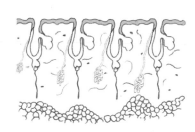

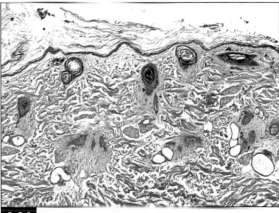

2-2c The lesions typically involve the hair shafts or the hair follicles

Congenital/hereditary hypotrichosis (e.g., Mexican hairless dog, Canadian hairless cat)

Endocrine disease

Feline paraneoplastic alopecia

Follicular dysplasia (e.g., color dilution alopecia)

Follicular infection

Glucocorticoid therapy

Mural folliculitis

Reduced vascular supply (ischemic lesions such as dermatomyositis)

Scarring that destroys adnexa

Systemic illness or therapy (telogen effluvium or anagen defluxion)

2-2d Note inactive hair follicles and lack of hair shafts in follicles.

continued

Table 2-1 Continued

CLINICAL LESION DEFINITION	RECOMMENDED METHOD OF BIOPSY SAMPLING IS ILLUSTRATED

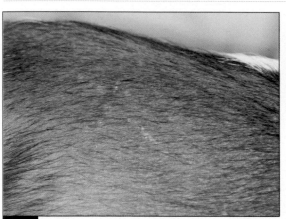 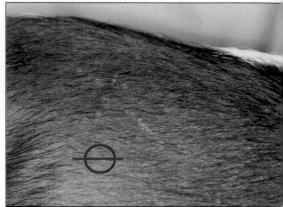

2-3a **Hypotrichosis:** partial hair loss (thinning of hair). Photograph reprinted with permission from Blackwell Publishing. Veterinary Dermatology 1991; 2: 69-83.

2-3b Punch biopsy samples of most extensively alopecic areas

Punch biopsy samples of normal areas labeled and kept separate from the partially haired areas

Punch biopsy samples; however, variety of techniques acceptable

Note a line is drawn through sample to illustrate to laboratory personnel how to cut sample

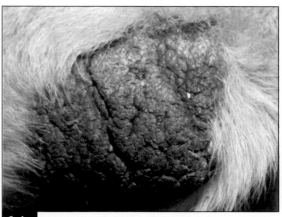 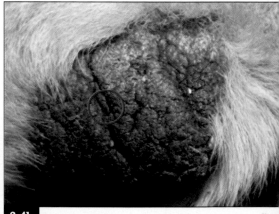

2-4a **Callus:** a thick hard hairless plaque with accentuation of superficial skin architecture (creases). (Courtesy of Dr. Ralf Mueller.)

2-4b Punch biopsy samples; however, variety of techniques acceptable

SKETCH OF LESION AND COMMON EXAMPLES

PHOTOMICROGRAPH OF EXAMPLE LESION

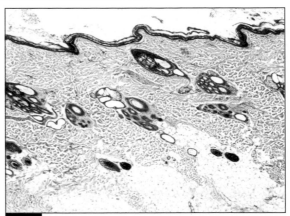

2-3c The lesions typically involve the hair shafts or follicles as for alopecia.

Endocrine disease

Follicular dysplasia (e.g., color dilution alopecia)

Follicular infection

Glucocorticoid therapy

Sebaceous adenitis (mechanism of hair loss unknown)

Self trauma

Systemic illness or therapy (telogen effluvium or anagen defluxion)

2-3d Note some hair follicles are inactive and others active. Some follicles have hair shafts, whereas others do not.

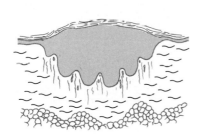

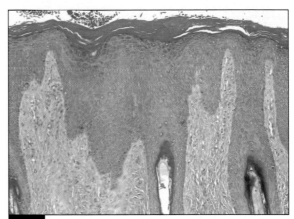

2-4c The lesion involves the stratum corneum and cellular layers of the epidermis. There may be follicular plugging leading to cysts.

Trauma over bony prominence such as elbow

2-4d Note epidermal hyperplasia and thickened stratum corneum.

continued

Table 2-1 Continued

CLINICAL LESION DEFINITION	RECOMMENDED METHOD OF BIOPSY SAMPLING IS ILLUSTRATED
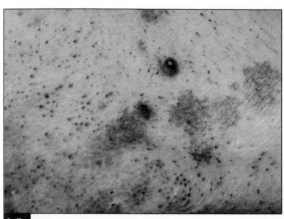 **2-5a** **Comedo:** plug of stratum corneum and sebum in a hair follicle. (Courtesy of Dr. Ralf Mueller.)	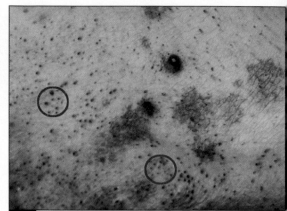 **2-5b** Punch biopsy samples; however, variety of techniques acceptable.
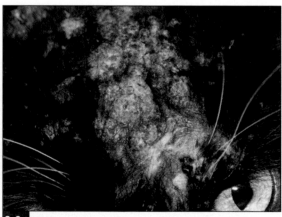 **2-6a** **Crust:** material formed by drying of exudate or secretion on the skin surface	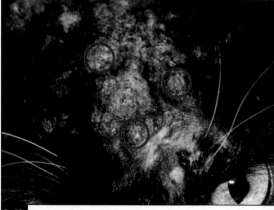 **2-6b** Punch biopsy samples; however, variety of techniques acceptable In addition, remove crust from other sites, and enfold crust in lens paper to keep crust intact. Place in formalin.

SKETCH OF LESION AND COMMON EXAMPLES

PHOTOMICROGRAPH OF EXAMPLE LESION

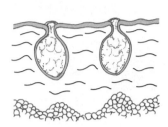

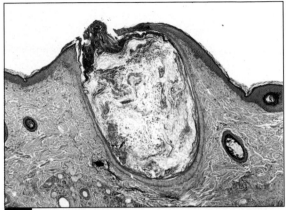

2-5c The lesion involves hair follicles.
Canine actinic dermatitis
Chin acne
Cornification disorders
Endocrine disorders
Schnauzer comedo syndrome
Secondary
   Surface trauma (e.g., callus)
   Demodicosis

2-5d Note cystic dilation of hair follicle with plug of stratum corneum at follicular opening.

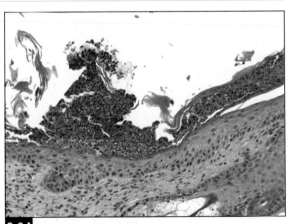

2-6c The lesion is on the surface of the epidermis (or surface of ulcer).
Advanced lesion in pustular diseases
   dermatophyte infection
   pemphigus foliaceus
   staphylococcal infection
Secondary to
   surface trauma
   lesions with fluid and cellular exudation

2-6d Note cellular debris in stratum corneum attached to surface of epidermis.

continued

Table 2-1 Continued

CLINICAL LESION DEFINITION	RECOMMENDED METHOD OF BIOPSY SAMPLING IS ILLUSTRATED
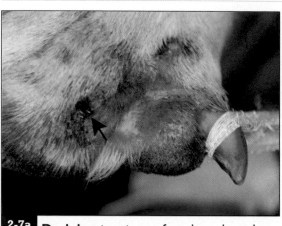**2-7a** **Draining tract:** an often deep dermal or subcutaneous nodular lesion in which exudate or necrotic debris dissects through the dermis and epidermis and drains onto the skin surface	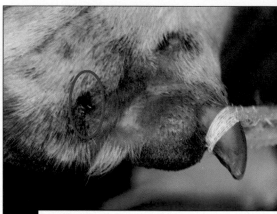**2-7b** Excisional, wide and deep enough to reach the complete lesion
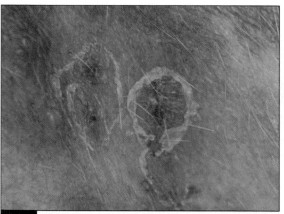**2-8a** **Epidermal collarette:** a ring of scale that expands peripherally	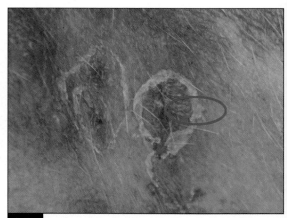**2-8b** Variety of sampling techniques possible. Include edge of scale attached to sample, and wrap sample in lens paper. Incisional (ellipse or wedge) or excisional samples may allow greater potential to identify cause.

SKETCH OF LESION AND COMMON EXAMPLES

PHOTOMICROGRAPH OF EXAMPLE LESION

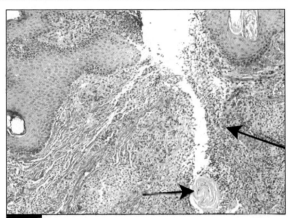

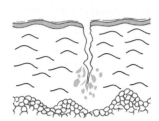

2-7c The primary lesion is in the dermis or subcutis.

    Deep bacterial infection
    Deep fungal infection
    Foreign body reaction

2-7d Arrows illustrate tract and "free" cornified cells released from ruptured follicle.

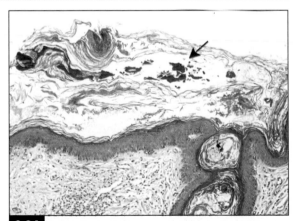

2-8c The lesion involves the stratum corneum and the epidermis.

Most common cause: superficial staphylococcal infection

Less common causes: fungal infection, neoplastic lesion, insect bite reaction, contact reaction

2-8d Note basophilic material (pustular debris) (arrow) within stratum corneum. This basophilic (bluish) pustular debris resulted in the name "blue line pyoderma."

continued

Table 2-1 Continued

CLINICAL LESION DEFINITION	RECOMMENDED METHOD OF BIOPSY SAMPLING IS ILLUSTRATED

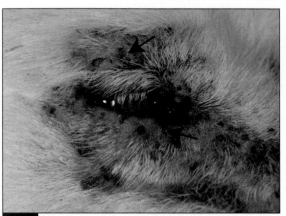 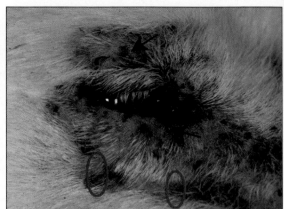

2-9a **Erosion:** loss of a portion of the epidermis that leaves the basement membrane intact

2-9b Variety of sampling techniques possible

Incisional (ellipse or wedge) or excisional samples may allow greater potential to identify cause

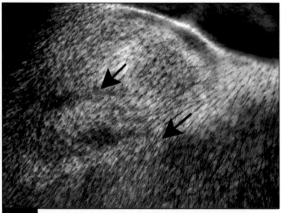 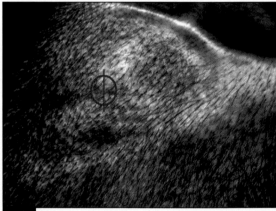

2-10a **Excoriation:** loss of epithelium due to physical trauma (scratching)

2-10b Punch biopsy samples; however, variety of techniques acceptable. Note a line is drawn through the sample to illustrate to laboratory personnel how to cut the sample.

SKETCH OF LESION AND COMMON EXAMPLES

PHOTOMICROGRAPH OF EXAMPLE LESION

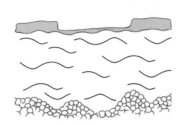

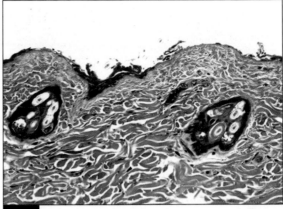

2-9c The lesion is within the epidermis.

Epidermal vesicular or bullous diseases

Pemphigus foliaceus

Superficial bacterial infection

Surface trauma

2-9d The surface is multifocally eroded. Note an occasional basal cell remains on the surface.

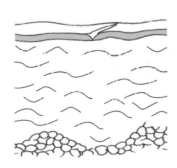

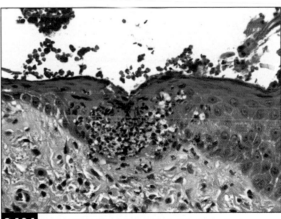

2-10c The lesion is within the epidermis, or if severe may extend into superficial dermis.

Surface trauma

2-10d Note small epidermal defect filled with neutrophils.

continued

Table 2-1 Continued

CLINICAL LESION DEFINITION	RECOMMENDED METHOD OF BIOPSY SAMPLING IS ILLUSTRATED
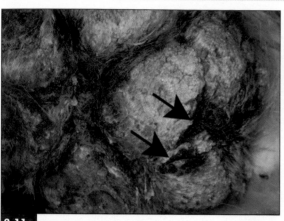 **2-11a** **Fissure:** linear defect within epidermis or through epidermis into dermis	 **2-11b** Incisional or excisional (if it could be curative)
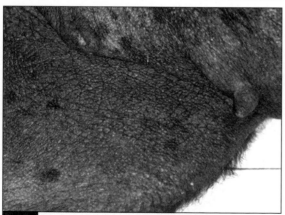 **2-12a** **Hyperpigmentation:** increase in epidermal and/or dermal melanin pigment	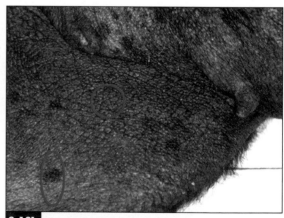 **2-12b** If diffuse, punch biopsy samples If localized, incisional (ellipse or wedge) or excisional samples may allow greater potential to identify cause

SKETCH OF LESION AND COMMON EXAMPLES

PHOTOMICROGRAPH OF EXAMPLE LESION

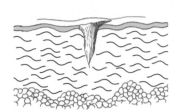

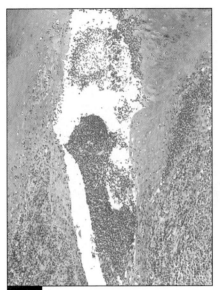

2-11c The lesion involves the epidermis and may extend into the dermis.

Footpad fissure may be associated with pemphigus foliaceus or superficial necrolytic dermatitis (metabolic epidermal necrosis)*

Nasal or digital hyperkeratosis

Thickened, frequently painful, hard skin with reduced resilience

2-11d Note fissure containing inflammatory debris in skin.

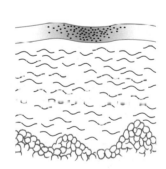

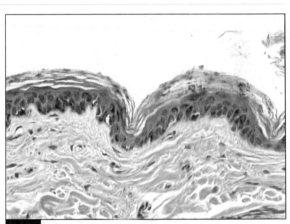

2-12c The lesion is within the epidermis but may also affect the superficial dermis.

Diffuse	Endocrine disease
Localized	Lentigo (see macule)
	Center of epidermal collarette
	Secondary to
	inflammation/injury
	hypersensitivity

2-12d Note melanin pigment in cellular epidermis and stratum corneum. It is difficult to diagnose hyperpigmentation without the knowledge of the normal amount of pigment expected for the site/breed (best determined by pigment in adjacent unaffected skin).

continued

* hepatocutaneous syndrome, diabetic dermatopathy, necrolytic migratory erythema

Table 2-1 Continued

CLINICAL LESION DEFINITION	RECOMMENDED METHOD OF BIOPSY SAMPLING IS ILLUSTRATED
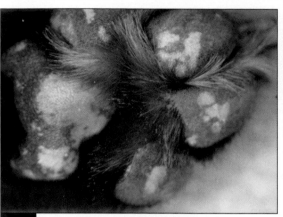 **2-13a** **Hypopigmentation:** a decrease in epidermal and/or dermal melanin pigment. (Courtesy of Dr. Michael Shipstone.)	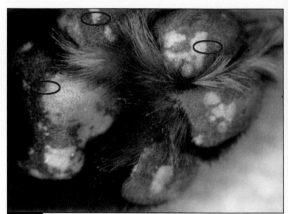 **2-13b** If diffuse, punch biopsy samples If localized, incisional (ellipse or wedge) or excisional samples may allow greater potential to identify cause
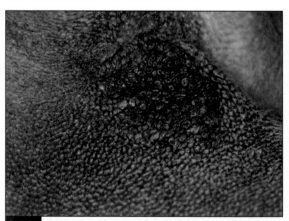 **2-14a** **Lichenification:** thickening of the skin with accentuation of superficial skin architecture (creases)	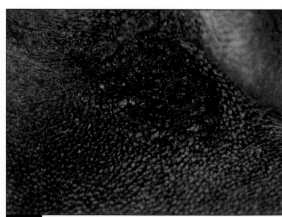 **2-14b** Punch biopsy samples; however, variety of techniques acceptable

SKETCH OF LESION AND COMMON EXAMPLES

PHOTOMICROGRAPH OF EXAMPLE LESION

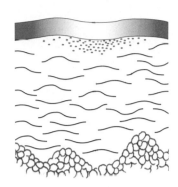

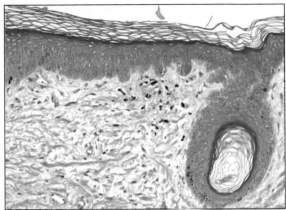

2-13c The lesion primarily involves the epidermis, but there may be lesions in the superficial dermis.

Dermatomyositis

Epitheliotropic lymphoma

Lupus erythematosus

Post inflammatory

Uveodermatologic syndrome (Vogt-Koyanagi-Harada-like syndrome)

Vitiligo

2-13d Most pigment has been lost from the epidermis and has been phagocytized by dermal macrophages. Note that epidermis has minimal pigment, and that the pigment is now in the dermis.

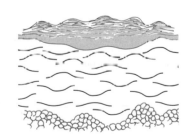

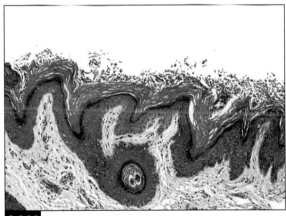

2-14c The lesion is in the epidermis and superficial dermis.

Acanthosis nigricans

Chronic infections

Chronic pruritic dermatoses
  allergies
  parasites

2-14d Note irregular epidermis that represents accentuation of superficial skin architecture.

continued

Table 2-1 Continued

CLINICAL LESION DEFINITION	RECOMMENDED METHOD OF BIOPSY SAMPLING IS ILLUSTRATED

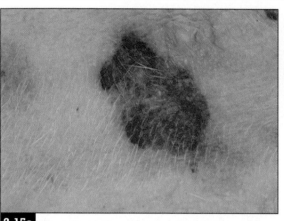 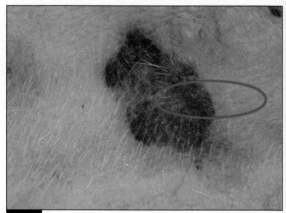

2-15a **Macule:** a flat circumscribed lesion of altered skin color (< 1cm)

2-15b Punch biopsy samples

Incisional (ellipse or wedge) or excisional samples may allow greater potential to identify cause

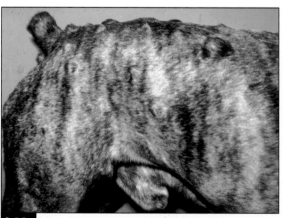 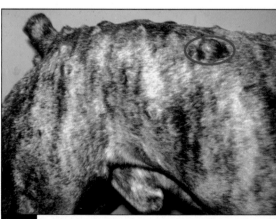

2-16a **Nodule:** a circumscribed, solid elevation of skin (> 1 cm). (Courtesy of Dr. Ralf Mueller.)

2-16b Excisional sample

SKETCH OF LESION AND COMMON EXAMPLES

PHOTOMICROGRAPH OF EXAMPLE LESION

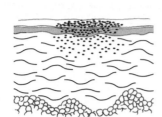

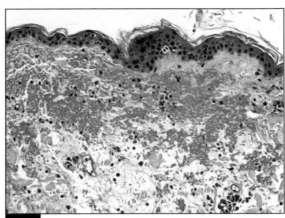

2-15c The lesion is in the epidermis or superficial dermis

Hemorrhage (redness)

Lentigo (increased pigment)

Vitiligo (reduced pigment)

2-15d Note dermal hemorrhage

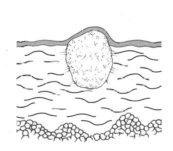

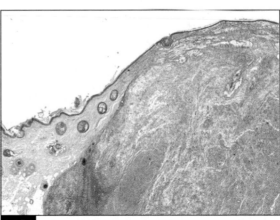

2-16c The lesion may develop in the epidermis, dermis, adnexa, or subcutis

Bacterial or fungal infection

Calcinosis circumscripta

Cyst

Sterile granuloma

Tumor

2-16d This photomicrograph illustrates the edge of the nodule. The nodule is seen as a mass of tissue, effacing normal architecture

continued

Table 2-1 Continued

CLINICAL LESION DEFINITION	RECOMMENDED METHOD OF BIOPSY SAMPLING IS ILLUSTRATED

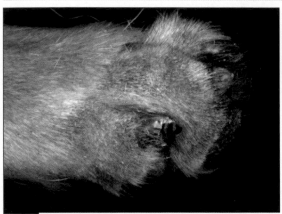 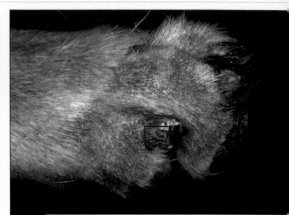

2-17a **Onychodystrophy:** claw deformity (abnormal formation of claw). Claw deformity may develop in association with paronychia or may be the only lesion. (Courtesy of Dr. Ralf Mueller.)

2-17b Claw bed punch biopsy technique (see Section I) or digital amputation can be used to diagnose lesions at the claw bed. The claw bed punch biopsy technique is not adequate for evaluating lesions in underlying bone.

Digital amputation is preferred for swelling or lysis of bone, which suggest secondary infection may have developed or there is a more serious underlying problem.

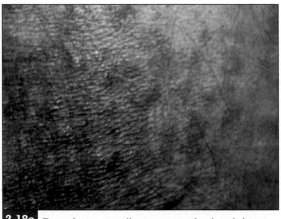 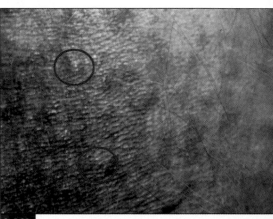

2-18a **Papule:** a small, circumscribed, solid elevation of skin (< I cm). Copyright Teton NewMedia Dermatology Made Easy Series.

2-18b Punch biopsy samples; however, variety of techniques acceptable

SKETCH OF LESION AND COMMON EXAMPLES

PHOTOMICROGRAPH OF EXAMPLE LESION

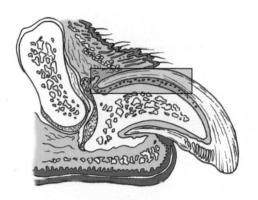

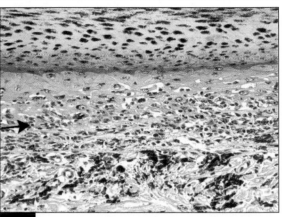

2-17c The lesion develops at the epidermal dermal junction of the claw bed or occasionally the vascular supply of the claw.

Idiopathic onychodystrophy

Infection (bacterial, fungal)*

Lupoid onychodystrophy

Lupus erythematosus*

Pemphigus (various forms)*

Vascular disease*

2-17d The lymphocytic inflammation obscures the epidermal dermal interface (arrow). Pigmentary incontinence is also present.

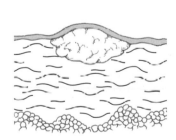

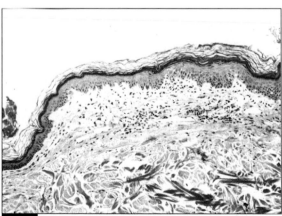

2-18c The lesion may develop in epidermis, dermis, or adnexa.

Insect bite

Small dermal mass

Small papilloma

Superficial folliculitis

Early pemphigus foliaceus

Hypersensitivity reactions

2-18d Note focus of mild inflammation causing localized elevation of epidermis.

continued

*Sampling of claw beds is not necessary if lesions are present in other more easily sampled sites.

Table 2-1 Continued

CLINICAL LESION DEFINITION	RECOMMENDED METHOD OF BIOPSY SAMPLING IS ILLUSTRATED

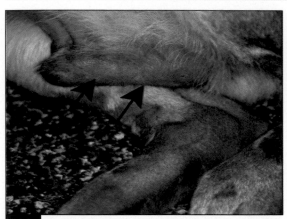

2-19a **Plaque:** a flat topped, solid elevation in the skin (> 1cm diameter) that occupies a relatively large surface area in comparison with its height

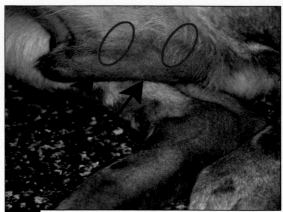

2-19b Punch samples collected within the plaque but avoiding the center where necrosis may be present

Incisional (ellipse or wedge) or excisional samples may allow greater potential to identify cause

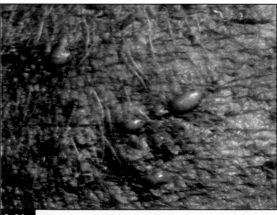

2-20a **Pustule, non follicular:** circumscribed accumulation of pus within the epidermis. (Courtesy of Dr. Rod Rosychuk.)

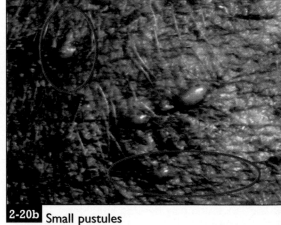

2-20b Small pustules
   Punch biopsy samples with pustule in center
Large pustules
   Excisional samples

SKETCH OF LESION AND COMMON EXAMPLES

PHOTOMICROGRAPH OF EXAMPLE LESION

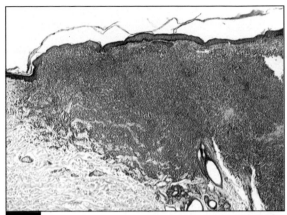

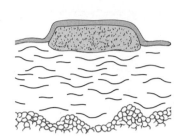

2-19c The lesion typically develops in the epidermis and/or dermis, but may involve adnexa.

Calcinosis cutis

Canine reactive histiocytosis

Coalescent papules *e.g., Demodex canis*

Eosinophilic plaque

Neoplasia

2-19d Note flat topped solid elevation of skin lesion.

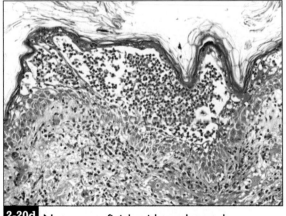

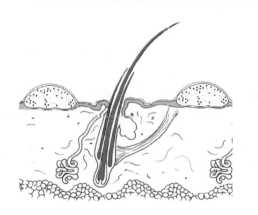

2-20c The lesion develops within the epidermis.

Bacterial or fungal disease

Sterile epidermal disease
    pemphigus foliaceus
    subcorneal pustular dermatosis

2-20d Note superficial epidermal pustule.

continued

Table 2-1 Continued

CLINICAL LESION DEFINITION	RECOMMENDED METHOD OF BIOPSY SAMPLING IS ILLUSTRATED

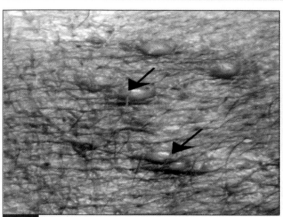**2-21a** **Pustule, follicular:** a pustule involving a hair follicle, typically a hair is visible emerging from the pustule. (Courtesy of Dr. Rod Rosychuk.)	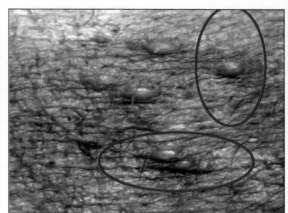**2-21b** Small pustules
	  Punch biopsy samples
	
	Large pustules
	  Excisional samples

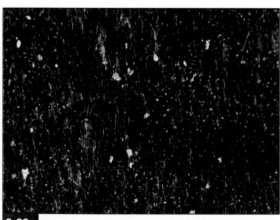**2-22a** **Scale:** thin and plate-like loose fragment of cornified cells resting on the surface of skin	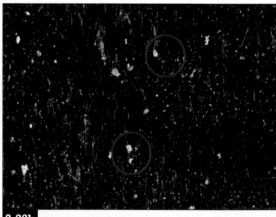**2-22b** Punch biopsy samples; however, variety of techniques acceptable Take care to include surface scale. May need to wrap sample in lens paper

SKETCH OF LESION AND COMMON EXAMPLES

PHOTOMICROGRAPH OF EXAMPLE LESION

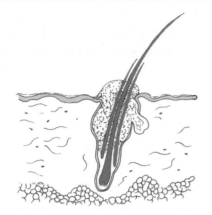

2-21c The lesion develops within hair follicles.

Demodicosis

Dermatophytosis

Pemphigus foliaceus is uncommon

Staphylococcal infection is the most common cause

2-21d Note exudate within lumen of follicle.

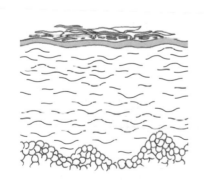

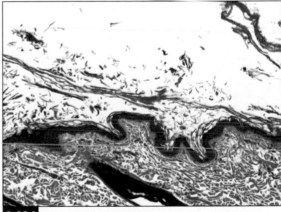

2-22c The lesion develops on the surface of the epidermis.

Cornification or keratinization disorders

Ichthyosis

Sebaceous adenitis

Secondary infections, allergies

Epitheliotropic lymphoma

Low environmental humidity

2-22d Note stratum corneum in fragments and sheets on surface and disproportionate increase in the stratum corneum.

continued

Table 2-1 Continued

CLINICAL LESION DEFINITION	RECOMMENDED METHOD OF BIOPSY SAMPLING IS ILLUSTRATED

2-23a **Scar:** an area of fibrous tissue that replaces components of epidermis, dermis, adnexa, and/or subcutis

May be raised (proliferative) or depressed (contracted)

2-23b Punch biopsy samples

Incisional (ellipse or wedge) or excisional samples may allow greater potential to identify cause.

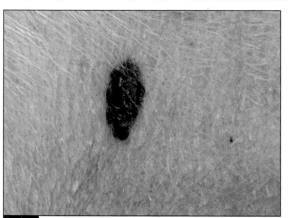

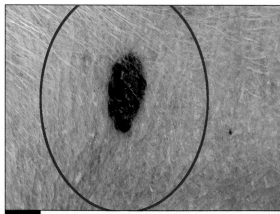

2-24a **Tumor:** "an abnormal mass of tissue, the growth of which exceeds and is uncoordinated with that of normal tissue and persists in the same excessive manner after cessation of the stimuli which evoked the change" (Willis, 1948)

2-24b Excision with 3 cm margins if malignant or invasive neoplasm is suspected.

SKETCH OF LESION AND COMMON EXAMPLES

PHOTOMICROGRAPH OF EXAMPLE LESION

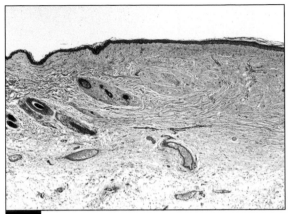

2-23c The lesion develops within the dermis.

Secondary to ischemic vascular injury
(e.g., dermatomyositis, vaccine-induced ischemic dermatitis, vasculitis in Jack Russell terriers)

Secondary to severe inflammatory diseases
(deep pyoderma, panniculitis, cellulitis)

Trauma (laceration, burn, chronic solar dermatitis)

2-23d Note adnexa are replaced by dense collagen fibers (scar). The skin surface is smoother than expected.

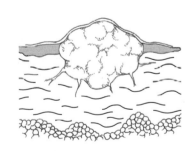

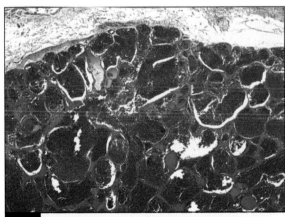

2-24c The lesion may develop in the epidermis, dermis, adnexa, or subcutis

See Section 4 for discussion of tumors

2-24d Note dermal nodule of blood filled vascular channels (hemangioma).

continued

Table 2-1 Continued

CLINICAL LESION DEFINITION	RECOMMENDED METHOD OF BIOPSY SAMPLING IS ILLUSTRATED

2-25a **Ulcer:** loss of the epidermis and basement membrane	**2-25b** Incisional (ellipse or wedge) or excisional samples may allow greater potential to identify cause
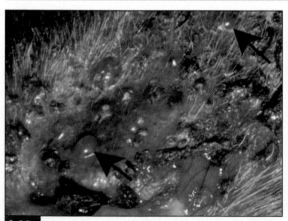	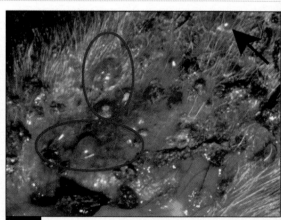
2-26a **Vesicle:** (blister) a localized collection of clear fluid usually in or beneath epidermis (arrows) (< 1.0cm). **Bulla:** a large vesicle (> 1.0cm) (Courtesy of Dr. Rod Rosychuk.)	**2-26b** Excisional, as most vesicles are fragile and require excisional sampling. A small vesicle may be excised with a large punch if the diameter of the vesicle clearly fits within the diameter of the punch instrument (2 mm around the edge of the vesicle)

SKETCH OF LESION AND COMMON EXAMPLES

PHOTOMICROGRAPH OF EXAMPLE LESION

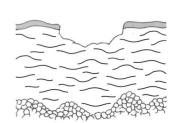

2-25c The lesion involves the loss of epidermis and basement membrane, and exposes superficial dermis

Feline herpes or calicivirus dermatitis

Feline indolent ulcer

Feline ulcerative dermatosis syndrome

Secondary to numerous inflammatory and immune mediated diseases

Subepidermal bullous diseases in which epidermis is lost

Trauma

Tumor (e.g., squamous cell carcinoma)

Vasculitis

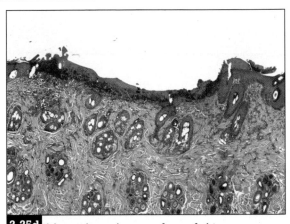

2-25d Note ulcer along surface of skin. The dark bluish purple material is necrotic superficial dermis.

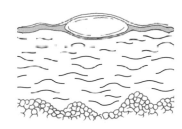

2-26c The lesion is located within or below the epidermis.

Adverse reaction to drug therapy

Congenital epidermolysis bullosa

Immune mediated vesicular and bullous diseases

Thermal or chemical burn

Viral infection herpes and calicivirus

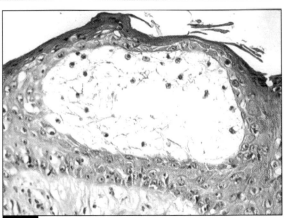

2-26d Note vesicle in epidermis.

continued

Table 2-1 Continued

CLINICAL LESION DEFINITION	RECOMMENDED METHOD OF BIOPSY SAMPLING IS ILLUSTRATED
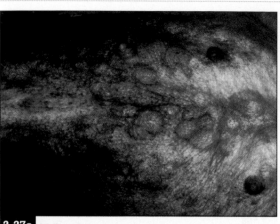 **2-27a** **Wheal:** a transient circumscribed raised edematous lesion (hive) (Courtesy of Dr. Douglas DeBoer.)	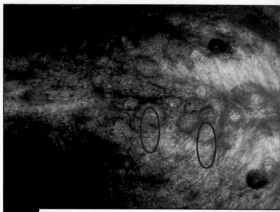 **2-27b** Incisional (ellipse or wedge) is ideal as it includes normal and abnormal skin, which helps facilitate diagnosis in subtle lesions Alternatively, several punch samples completely within lesional (wheal area) plus several samples completely within adjacent normal skin would be acceptable.

SKETCH OF LESION AND COMMON EXAMPLES

PHOTOMICROGRAPH OF EXAMPLE LESION

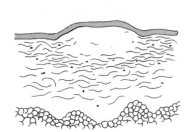

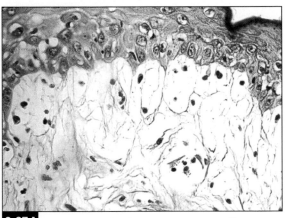

2-27c The lesion is located within the dermis and causes elevation of the epidermis. The epidermis may be thickened by acanthosis and edema.

Hypersensitivity (allergic) response
    Drug reaction (topical, oral, injectable)
    Insect bite
    Other
    Cold, physical trauma

2-27d Note dermal edema separating collagen fibers.

Table 2-2
Secondary Clinical Lesions Associated with Surface Trauma*

Secondary lesions associated with biting, chewing, and scratching**	Secondary lesions associated with licking or rubbing**	Secondary lesions associated with trauma from hard flooring
Excoriation	Lichenification	Callus
Pustules, crusts, scales (in part due to secondary bacterial infection)	Callus	Secondary follicular plugging, cyst formation, cyst rupture (furuncle), and draining tract (callus pyoderma)
Erosions and ulcers	Erosion/ulceration	
Hypotrichosis and alopecia	Hyperpigmentation	Fibroadnexal hamartoma (synonym: focal adnexal dysplasia)
Scarring	Alopecia	
	Follicular plugging with follicular cyst formation and furunculosis (can cause draining nodular lesion)	
	Scarring (secondary to ulceration or follicular rupture)	

*Surface trauma can be caused by pruritus, boredom, psychogenic problems, or environmental conditions. A search for primary lesions (e.g., papules) should be performed in an attempt to identify the underlying basis for the lesions (e.g., hypersensitivity, superficial bacterial or yeast infection, etc.). The inflammation associated with these secondary lesions can mask the primary disease process and make it impossible for the pathologist to identify.
**These lesions are frequently present in the same animal.

Section 3

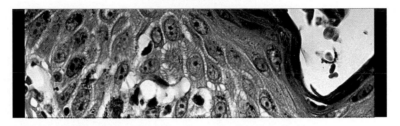

Histopathologic Responses
of the Skin to Injury

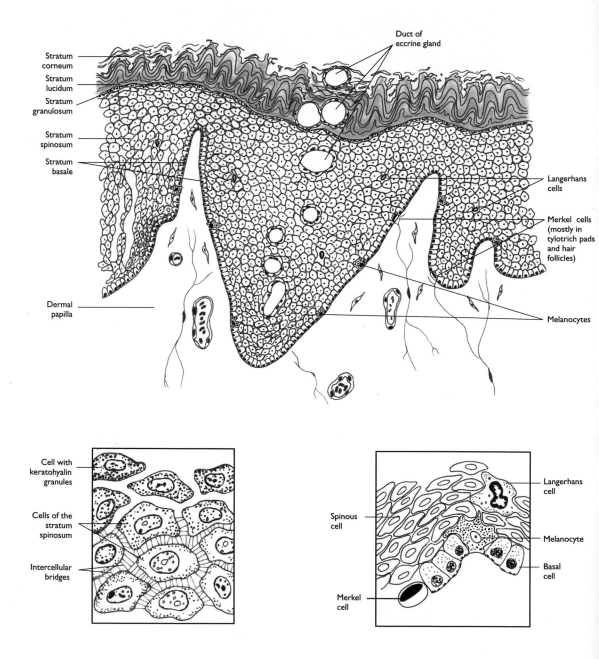

3-1 Epidermis, footpad skin

Histopathologic Responses of the Epidermis to Injury

✔ Before discussing the responses of the epidermis to injury, a review of epidermal histology is helpful (Figure 3-1). Where skin is haired, the epidermis has four layers or strata. These layers include, from the innermost to the outermost, the stratum basale, spinosum, granulosum, and corneum. Where skin is not haired (*e.g.*, footpad and planum nasale), there may be an additional layer or strata called, the stratum lucidum, which is located between the stratum corneum and stratum granulosum. The cells within the stratum basale are the germinal cells that undergo mitosis and produce epidermal cells. The cells in the stratum basale also anchor the epidermis to the dermis via hemidesmosomes. Cells within the stratum spinosum and granulosum are maturing cells that produce keratin filaments and keratohyalin granules. The cells within the stratum corneum are composed of a compacted layer of anuclear squamous cells consisting in part of keratin filaments and keratohyalin granules. The cells within these epidermal layers are referred to as keratinocytes and they comprise the majority of the epidermis (about 85%). Keratinocytes are attached to each other by desmosomes.

✔ Also located within the epidermis, are melanocytes (produce melanin pigment), Langerhans cells (process and present antigen to sensitized T-lymphocytes), and Merkel cells (may have variety of functions including serving as mechanoreceptors, and in a paracrine role by regulating the function of adjacent epidermal and adnexal structures). Melanocytes and Merkel cells are in the basal layer, but Langerhans cells may be in any layer. The epidermis and dermis are separated by a basement membrane zone that consists of the basal cell plasma membrane, the lamina lucida, the lamina densa, and the sublamina densa. This zone is composed of a matrix of glycoproteins and macromolecules. Defects or damage to the adhesion structures of the epidermis or basement membrane zone (*e.g.*, desmosomal or hemidesmosomal proteins) or other components of the basement membrane (*e.g.*, antigens, collagen, laminin) may lead to acantholytic and bullous diseases. In haired skin, the epidermis is about three cells in thickness and the basement membrane zone is smooth because the epidermis and dermis are protected by the hair coat (hair follicles help serve to anchor the epidermis and dermis). However in footpads and nasal planum, the cellular layers increase (up to about 20 cells, mostly within the stratum spinosum) and the contour of the basement membrane zone is irregular due to epidermal dermal interdigitations that help strengthen the attachment of the epidermis to the dermis.

Tables 3-1 through 3-4 review the responses of the epidermis to injury.

Table 3-1 Alterations in Epidermal Growth or Differentiation

RESPONSE OF THE EPIDERMIS TO INJURY	HISTOLOGIC FEATURE	SALIENT CLINICAL FEATURE

Hyperkeratosis: histologic term for thickening of stratum corneum (cornified layer)
 	
Synonyms
 	
Cornification disorder = abnormal production of the keratin, lipids, or other components of this layer
 	
Keratinization disorder = abnormal production of the keratin proteins
 	
Cornification and keratinization are terms that although used interchangeably, are actually different processes
	Increased thickness of the layers of stratum corneum
	
	Orthokeratotic hyperkeratosis— no nuclei retained in the stratum corneum
	Subtypes include
	  basket weave (loosely woven)
	  compact (compressed, dense)
	  laminated (plate-like thickening)
	
	Parakeratotic hyperkeratosis— nuclei retained within the stratum corneum
		Orthokeratotic hyperkeratosis: Loose scales, forming "dandruff"
		
		Parakeratotic hyperkeratosis: Thickened skin with tightly adherent scales
		
		There may be dried fluid and cellular debris mixed with the parakeratotic cornified cells forming crusts

DISEASE EXAMPLES

Hyperkeratosis

Actinic keratosis

Cutaneous horn

Ear margin seborrhea

Epidermolytic hyperkeratosis (Figure 3-2)

Digital hyperkeratosis, familial

Ichthyosis (Figure 3-3)

Naso-digital hyperkeratosis
  Idiopathic
  Canine distemper

Primary idiopathic seborrhea

Sebaceous adenitis

Vitamin A responsive dermatosis

Parakeratosis

Zinc responsive dermatosis (Figure 3-4)

Hereditary nasal parakeratosis in Labrador retrievers

Acrodermatitis of Bull terriers

Superficial necrolytic dermatitis (metabolic epidermal necrosis)* (Figure 3-5)

Superficial suppurative necrolytic dermatitis of miniature Schnauzers

Both

Common secondary change seen with chronic inflammation or surface trauma

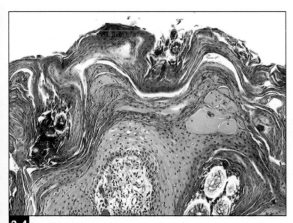

3-4 Parakeratotic hyperkeratosis in a Husky with zinc responsive dermatosis.

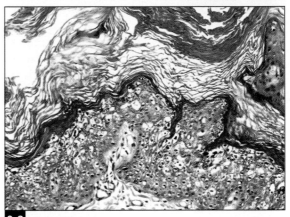

3-2 Orthokeratotic hyperkeratosis, vacuolation of cytoplasm of epidermal cells in the granular layer, and increased variation of granule size from a Rhodesian ridgeback with epidermolytic hyperkeratosis.

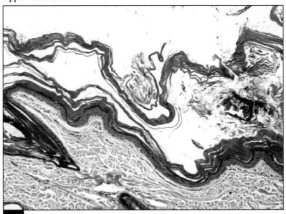

3-3 Compact hyperkeratosis in a Golden retriever with ichthyosis. Note the sheets of stratum corneum on the surface of the epidermis.

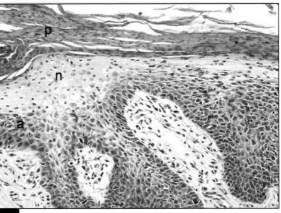

3-5 Parakeratosis (p), subparakeratotic pale zone (necrolysis) (n), and epidermal acanthosis (a) providing a trilaminar appearance to the epidermis (red, white, and blue), in a dog with superficial necrolytic dermatitis.

continued

* Additional synonyms include (hepatocutaneous syndrome, diabetic dermatopathy, necrolytic migratory erythema, metabolic epidermal necrosis)

Table 3-1 Continued

REACTION OF THE EPIDERMIS TO INJURY	HISTOLOGIC FEATURE	SALIENT CLINICAL FEATURE
Acanthosis (hyperplasia): thickening of the spinous cell layer (stratum spinosum)	The cellular portion of the epidermis is thickened by increased numbers of spinous layer cells Types Psoriasiform—elongated but regular epidermal dermal interdigitations Papillated—epidermis that is acanthotic and also forms finger like projections above the skin surface Pseudocarcinomatous—thickening of the epidermis with downward extensions of the epidermis into the dermis mimicking features of squamous cell carcinoma	Mild hyperplasia (acanthosis) is unlikely to be clinically apparent Moderate hyperplasia (acanthosis) causes accentuation of skin creases (lichenification) and is typically accompanied by increased pigmentation (hyperpigmentation) Extensive hyperplasia (acanthosis) results in marked epidermal thickening and irregularity, and is also apparent by palpation
Dyskeratosis: a confusing morphologic term generally used for premature or abnormal keratinization of individual cells within viable epidermis. Dyskeratosis is difficult to impossible to distinguish from apoptosis or necrosis by histopathologic examination.	Within epidermis, individual keratinocytes are rounded, detached from neighboring keratinocytes, and are more eosinophilic due to accumulation of keratin filaments	No significant gross alteration with mild dyskeratosis

DISEASE EXAMPLES

Acanthosis nigricans

Lichenoid psoriasiform dermatitis of Springer spaniel

Common secondary change seen with chronic inflammation or surface trauma

Examples
    Acral lick dermatitis
    (Figure 3-6)
    Callus
    Chronic allergic dermatitis
    (Figure 3-7)
    Edge of persistent ulcer

Zinc responsive dermatosis

3-6 Acanthosis in Doberman pinscher with chronic surface trauma from persistent licking. Note compact keratin, acanthosis, and hypergranulosis. There also is dermal scarring.

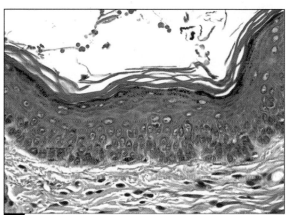

3-7 Acanthosis in dog with chronic atopic dermatitis. Thickening of the stratum spinosum (true definition of acanthosis) is seldom seen as the sole change. In this section there is concurrent compact orthokeratotic hyperkeratosis and an increased granular cell layer, both changes consistent with rubbing.

continued

Table 3-1 Continued

REACTION OF THE EPIDERMIS TO INJURY	HISTOLOGIC FEATURE	SALIENT CLINICAL FEATURE
Apoptosis: physiological or programmed cell death Apoptotic cells have the morphologic features of dyskeratotic and necrotic cells, the distinction must be made by electron microscopy Sometimes apoptotic cells are referred to as "individual necrotic cells," but the process of apoptosis is not the same as necrosis, thus apoptotic cells should not be referred to as necrotic cells.	Scattered individual keratinocytes are shrunken, rounded cells detached from neighboring keratinocytes. Apoptotic cells have increased eosinophilia due to condensation of organelles. A few lymphocytes may border these keratinocytes (satellitosis).	No significant gross alteration with apoptosis alone; however, if extensive, it may result in erosion or ulceration
Necrosis: death of epidermal cells in the living animal*	Loss of nuclear and cytoplasmic detail. Nucleus may be shrunken (pyknotic), fragmented (karyorrhectic), or missing (karyolytic).	Gross lesions vary with extent and depth of the necrosis. The epidermis may be separated from dermis by vesicle fluid or may be dislodged easily from the dermis with digital pressure.
Of the different types of necrosis, coagulation necrosis affects the epidermis.	The cell is homogeneous and deeply eosinophilic (taking pink/red dye), and the cell outline persists after cell detail has disappeared.	Necrotic tissue is sharply delineated from adjacent skin. Dry thick epidermis can be lifted from the dermis leaving full thickness ulcer.

* The degree of necrosis may vary affecting a few cells or confluent areas

DISEASE EXAMPLES

Adverse reaction to drugs

Dermatomyositis

Erythema multiforme
(Figure 3-8)**

Feline exfoliative dermatosis
associated with thymoma

Graft versus host disease

Lupus erythematosus (discoid,
systemic)

Lupus erythematosus, cutaneous
exfoliative (lupoid dermatosis of
the German shorthair pointer)

Lupus erythematosus, vesicular
(ulcerative dermatosis of the
Collie and Shetland sheepdog)

Neoplasia (squamous cell
carcinoma (Figure 3-9)

Proliferative otitis of young cats

Sunburn

Vaccine-induced ischemic
dermatitis

Vasculitis in Jack Russell terriers

Normal, regulation of tissue
growth, and of particular
relevance to skin, the involution
of hair follicles associated with
changes in the hair cycle

Chemical or thermal burn
(Figure 3-10a)

Vasculitis or thrombosis
   any type of vasculitis or
   condition that may cause
   thrombosis (e.g., cyroglobulin
   disorder, proliferative
   thrombovascular necrosis of
   the pinnae, septic or sterile
   thrombi causing infarction)

Venomous bites (spider,
snake, other)

Frostbite

** There is overlap between erythema multiforme and toxic epidermal necrosis.

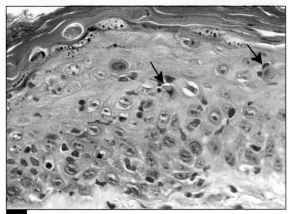

3-8 Apoptosis (physiological or programmed cell death) associated with immune mediated killing of epidermal cells in erythema multiforme. Note apoptotic keratinocytes and lymphocytes (satellitosis) bordering the apoptotic keratinocytes (arrows).

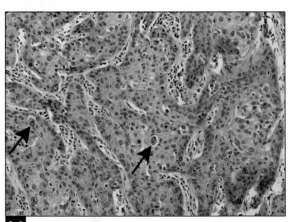

3-9 Apoptosis associated with neoplastic disease (squamous cell carcinoma). Note the eosinophilic keratinocytes (arrows).

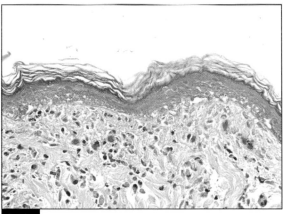

3-10a Epidermal necrosis associated with thermal burn. Note coagulation necrosis of epidermis.

continued

Table 3-1 Continued

REACTION OF THE EPIDERMIS TO INJURY	HISTOLOGIC FEATURE	SALIENT CLINICAL FEATURE

Necrolysis: separation of tissue due to death of cells
 	
Can occur at various levels within the epidermis or beneath epidermis.
	Toxic epidermal necrolysis-Stevens-Johnson syndrome or Thermal burn
	
	    the epidermis is necrotic and separated from dermis by vesicle fluid or may easily be dislodged from the dermis with digital pressure
		Toxic epidermal necrolysis-Stevens-Johnson syndrome
		    vesicular or bullous eruption leading to epidermal necrosis and slough
		
		Thermal burn
		    coagulation necrosis of the epidermis leading to separation of the epidermis from the dermis

	Superficial necrolytic dermatitis (metabolic epidermal necrosis)
	    thick parakeratotic stratum corneum (red) is separated by a zone of reticular degeneration (necrolysis) in superficial stratum spinosum (white) from underlying acanthotic stratum spinosum (blue)
		Superficial necrolytic dermatitis (metabolic epidermal necrosis)
		    pustular areas, crusts, ulcers, footpad crusting and fissuring, paronychia

	Superficial suppurative necrolytic dermatitis in miniature schnauzers
	    parakeratosis with superficial reticular degeneration (necrolysis), suppuration, and acanthosis
		Superficial suppurative necrolytic dermatitis in miniatures schnauzers
		    predominantly ventral papules, plaques, pustules, ulcers

Ulcer: loss of epidermis and the basement membrane	Depressed area due to absence of the epidermis and superficial dermis.	Focal depression of skin surface due to loss of epidermis and superficial dermis. The underlying dermis is exposed. Ulcers frequently are covered by serous or purulent exudate, and crusts.

DISEASE EXAMPLES

Toxic epidermal necrolysis-
Stevens-Johnson syndrome*
Thermal burn involving all layers
of the epidermis (Figure 3-10b)

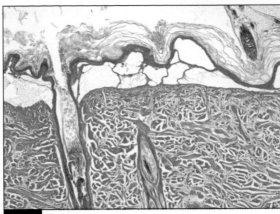

3-10b Thermal burn in Dalmation puppy. Note epidermis is separated from dermis by vesicular spaces. The vesicle fluid washed out during processing.

Superficial necrolytic dermatitis
(metabolic epidermal necrosis)
(Figure 3-10c)

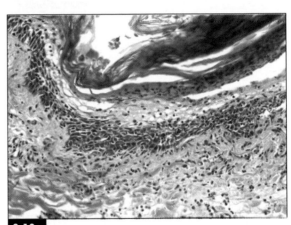

Superficial suppurative
necrolytic dermatitis in
Miniature schnauzers

3-10c Superficial necrolytic dermatitis (metabolic epidermal necrosis) in a dog. Note the red, white and blue appearance of the epidermis. There is mild inflammation in the epidermis and dermis.

Feline herpesvirus dermatitis

Feline indolent ulcer

Feline ulcerative dermatosis
syndrome

Mosquito bite hypersensitivity

Secondary to epidermal necrosis

Secondary to trauma
(Figure 3-11)

Secondary to vesicular/bullous
diseases (dermatomyosis
erythema multiforme)

Tumor

Vasculitis (typically punctate ulcers)

Vesicular lupus erythematosus
(ulcerative dermatosis
syndrome of the Collie and
Shetland sheepdog)

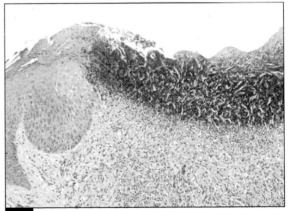

3-11 Ulcer from the skin of a cat with allergic dermatitis and self trauma. The ulcerated surface is covered by degenerate cellular debris.

continued

* There is overlap between erythema multiforme, Stevens-Johnson syndrome, and toxic epidermal necrolysis.

Table 3-1 Continued

REACTION OF THE EPIDERMIS TO INJURY	HISTOLOGIC FEATURE	SALIENT CLINICAL FEATURE
Atrophy: reduction in size of a cell, tissue, organ, or part	Thinning of the cellular layers of the epidermis. Atrophy may be difficult to evaluate as normal canine and feline epidermis are only two to three nucleated cell layers thick.	Usually there also is atrophy of dermis resulting in palpably thin skin through which vessels may be visualized.
Dysplasia: abnormality of development or organization	Variation from the normal organizational pattern or structure of the epidermis.	Varies with disease entity.

	Chronic moderate heat dermatitis
	  large keratinocyte nuclei, apoptotic basal cells and keratinocytes, eosinophilic elastic tissue in dermis
		Chronic moderate heat dermatitis
		  mottled or reticulated erythema and alopecia

	Preneoplastic changes
	  irregular stratification, variation in size and shape of cells, increase basaloid cells
		Preneoplastic lesions (e.g., Bowen's disease and actinic dermatitis)
		  scaly plaques

	Feline leukemia virus
	  multinucleated giant cells, disordered keratinocyte maturation, dyskeratosis, leading to necrosis and ulceration
		Feline leukemia virus
		  scaling, crusting, alopecia

DISEASE EXAMPLES

Feline skin fragility syndrome

Hyperadrenocorticism

Severe nutritional deprivation

Topical steroid associated
atrophy (Figure 3-12a)

Chronic moderate heat
dermatitis (erythema ab igne)

Preneoplastic lesions
    Bowen's-like disease
    (Figures 3-12b and 3-12c)
    Actinic dermatitis

Feline leukemia virus dermatitis

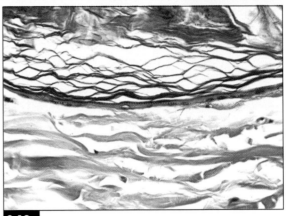

3-12a Atrophy of epidermis from ventral abdominal skin of a dog that has had long term therapy with topical glucocorticoids

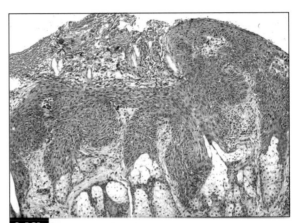

3-12b Dysplasia of epidermis in a cat with Bowen's-like disease. Note irregular stratification of epidermis and increased basloid cells.

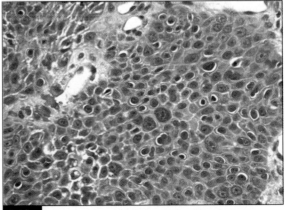

3-12c Dysplasia of epidermis in a cat with Bowen's-like disease. Note variation in nuclear size.

Table 3-2 Alterations in Epidermal Cell Adhesion

REACTION OF THE EPIDERMIS TO INJURY	HISTOLOGIC FEATURE	SALIENT CLINICAL FEATURE
Edema: fluid accumulation between epidermal cells (spongiosis) or within epidermal cells	Extracellular fluid results in widening of intercellular spaces resulting in accentuation of intercellular bridges (spongiosis) Intracellular edema results in swelling of cell cytoplasm*	A small vesicular eruption is the classic clinical lesion of spongiosis. The vesicles may rupture and crust, and are frequently associated with erythema.

Hydropic degeneration: intracellular edema of basal epidermal cells	Hydropic degeneration
	  basal layer cells have swollen
	  vacuolated cytoplasm	

Ballooning degeneration: intracellular edema resulting in swelling of cells and loss of intercellular attachments	Ballooning degeneration
	  swollen epidermal cells with
	  loss of intercellular
	  attachments	

Acantholysis: loss of attachment between keratinocytes due to the breakdown of intercellular bridges. May occur via immunologic mechanisms, neutrophilic enzymatic destruction, physical separation due to extracellular edema, damage to keratinocytes in viral infections, or genetic abnormalities	Epidermal cells may be organized irregularly and lose attachment to form intraepidermal vesicles with free floating epidermal cells (acantholytic cells). The acantholytic cells may occur singly or in groups (rafts). In pemphigus foliaceus, the pustules containing acantholytic cells rapidly crust, so the acantholytic cells are frequently embedded within serocellular crusts.	Papules and small vesicles are the earliest visible lesions Fully developed lesions of acantholysis result in formation of larger vesicles or pustules that quickly evolve to erosions, ulcers, or crusts.

* Intracellular edema, also called hydropic, vacuolar, or ballooning degeneration, causes cellular swelling. If the swelling is severe, the edematous cells may rupture and form microvesicles that are supported by the cell walls of the ruptured cells. This type of epidermal damage is called reticular degeneration. Intracellular edema limited to the basal layer of the epidermis is referred to as hydropic or vacuolar degeneration. Hydropic degeneration may lead to intra-basilar vesicles. Ballooning degeneration is intracellular edema of cells in more superficial layers of the epidermis. The swollen epidermal cells lose intercellular attachments and may form an intraepidermal vesicle. Ballooning degeneration is a feature of some viral diseases, particularly poxvirus.

DISEASE EXAMPLES

Edema: allergic contact dermatitis or other hypersensitivity reactions including insect bites, and severe inflammation associated with dermal edema that may spread into the epidermis (Figure 3-13)

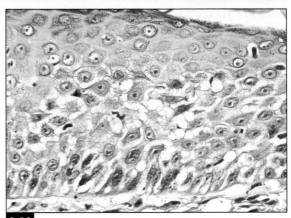

3-13 Extracellular edema of the epidermis in a dog with a cutaneous hypersensitivity reaction. Note widening of intercellular spaces illustrating intercellular bridges.

Hydropic degeneration: dermatomyositis, drug reactions, lupus erythematosus (Figure 3-14)

Ballooning degeneration: viral infection (e.g., poxvirus)

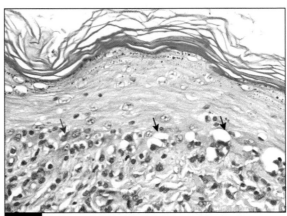

3-14 Interface dermatitis in Shetland sheepdog with lupus erythematosus. Note vacuolization of basal cells (hydropic degeneration) (black arrows) and occasional apoptotic basal cell (red arrow). Mild interface inflammation is also present.

Congenital acantholytic diseases (Darier's disease)

Pemphigus erythematosus

Pemphigus foliaceus (Figure 3-15)

Pemphigus vulgaris

May form in pustules associated with infectious agents (e.g., staphylococci or less frequently *Trichophyton* sp.)

Squamous cell carcinoma (acantholytic subtype)

Viral vesicles (e.g., herpesvirus)

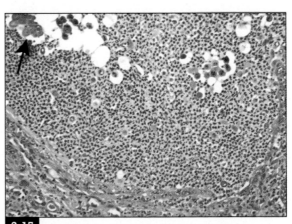

3-15 Acantholytic cells in pustule from a dog with pemphigus foliaceus. Note raft of acantholytic cells in pustule (arrow).
continued

Table 3-2 Continued

REACTION OF THE EPIDERMIS TO INJURY	HISTOLOGIC FEATURE	SALIENT CLINICAL FEATURE
Vesicle: small intraepidermal or subepidermal blister (< 1.0cm) **Bulla:** large vesicle (> 1.0cm)	Fluid-filled cavity within or below the epidermis further defined by the level at which it develops (intraspinous, suprabasilar, subepidermal)	Circumscribed raised area in or beneath the epidermis and filled with clear fluid Vesicular lesions are usually transient and rapidly lead to ulcers, a more common clinical feature. Ulcers typically progress to crusts.

DISEASE EXAMPLES

Burn, thermal or chemical

Congenital epidermolysis bullosa

Lupus erythematosus

Pemphigus vulgaris (Figure 3-16)

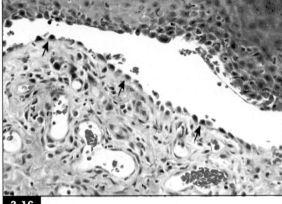

3-16 Suprabasilar vesicle in dog with pemphigus vulgaris. Note the row of basal cells remaining attached to the basement membrane (referred to as row of tombstones) (arrows).

Subepidermal bullous diseases
   Bullous pemphigoid (Figure 3-17)

   Mucous membrane pemphigoid

   Linear IgA dermatosis

   Epidermolysis bullosa acquisita

Toxic epidermal necrolysis

Viral infection (e.g., herpesvirus)

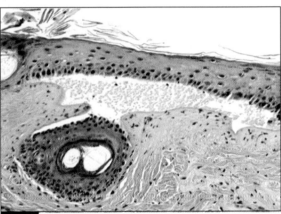

3-17 Subepidermal vesicle in Border collie with bullous pemphigoid.

Table 3-3 Inflammatory Epidermal Lesions

RESPONSE OF THE EPIDERMIS TO INJURY	HISTOLOGIC FEATURE	SALIENT CLINICAL FEATURE
Exocytosis: migration of leukocytes or erythrocytes between the cells of the epidermis (intercellular)	Leukocytes or erythrocytes are in the intercellular epidermis, and may be accompanied by intercellular edema	Mild exocytosis is not clinically evident The epidermis may appear thickened or puffy with significant amounts of spongiosis and exocytosis Petechial hemorrhage is a feature of severe erythrocytic exocytosis
Pustules: circumscribed accumulation of inflammatory cells within the epidermis	Intraepidermal vesicle filled with inflammatory cells. Location within epidermis and type of inflammatory cell are key features in determining the cause of the pustule **Bacterial** (intracorneal or epidermal, degenerate neutrophils, cocci) **Pemphigus foliaceus** (subcorneal or intraspinous, healthy neutrophils, acantholytic cells) **Parasitic hypersensitivity** (superficial epidermal, eosinophils). A small wedge shaped focus of epidermal degeneration associated with eosinophils is feature of a bite or nibble by a parasite. **Epitheliotropic lymphoma** (intraepidermal, lymphocytes)	Circumscribed raised area in the epidermis filled with leukocytes. Some pustules are too small to see grossly and are termed "micropustules."

DISEASE EXAMPLES

Many inflammatory conditions are associated with leukocytic exocytosis (Figures 3-18a and 3-18b)

Cutaneous infection: *Malassezia*, bacteria, dermatophytes

Hypersensitivity reactions: contact, insect bites, atopic dermatitis

Immune mediated disease: pemphigus foliaceus, erythematosus

Erythrocytic exocytosis is associated with trauma, vasculitis, or coagulopathies

Mild erythrocytic exocytosis may be seen with other conditions in which there is secondary damage to blood vessels

Eosinophilic pustules associated with parasitic hypersensitivity

Pautrier's microabscess associated with epitheliotropic lymphoma

Pemphigus erythematosus

Pemphigus foliaceus

Subcorneal pustular dermatosis

Superficial bacterial infection (impetigo) (Figure 3-19)

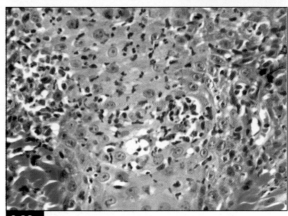

3-18a Leukocytic exocytosis in an atopic dog with superficial bacterial infection.

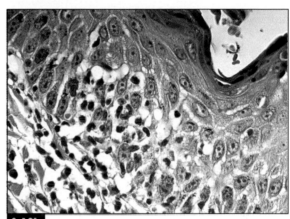

3-18b Leukocytic exocytosis in a dog with superficial infection with *Malassezia pachydermatis*. Focal lymphocytic exocytosis can provide a clue for infection with *Malassezia*, even in the absence of histologically detectable yeast. There is also spongiosis.

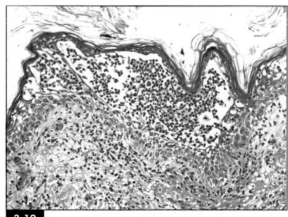

3-19 Intraepidermal pustule in a 6 month old puppy with impetigo.

continued

Table 3-3 Continued

RESPONSE OF THE EPIDERMIS TO INJURY	HISTOLOGIC FEATURE	SALIENT CLINICAL FEATURE
Crusts: material formed by drying of exudate or secretion on the skin surface	Variably thick coagulum of shrunken cells (inflammatory and epidermal), proteinaceous fluid, and debris on the skin surface Crusts may contain infectious agents or acantholytic cells that help provide diagnostic clues when examining histologic sections	Thick, irregular, dry, sometimes friable material that may be loosely attached or adherent to the skin surface

DISEASE EXAMPLES

Common to many diseases and may not be clinically diagnostic, but histologically may hold key to diagnosis (e.g., dermatophyte hyphae in *Trichophyton sp.* infection, acantholytic cells in pemphigus foliaceus) (Figure 3-20)

Examples
Erosive or ulcerative diseases (self trauma, furunculosis, vasculitis)

Immune mediated disease (especially pemphigus foliaceus, lupus erythematosus)

Superficial infections (bacterial, dermatophyte, yeast such as *Candida* sp.)

Excessive cornification with exudation (zinc responsive dermatosis)

Trauma

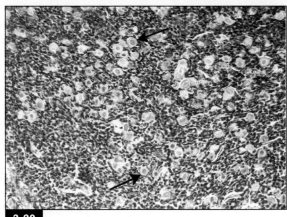

3-20 Crust in a dog with pemphigus foliaceus. Note the rounded, eosinophilic acantholytic cells (arrows).

Table 3-4 **Alterations in Epidermal Pigmentation**

RESPONSE OF THE EPIDERMIS TO INJURY	HISTOLOGIC FEATURE	SALIENT CLINICAL FEATURE
Hyperpigmentation: increased amount of melanin pigment in the epidermis or superficial dermis	Increased melanin pigment granules in the cytoplasm of keratinocytes, within macrophages and melanocytes in the perivascular dermis, or sometimes increased numbers of melanocytes. Knowledge of the amount of melanin pigment in the normal skin (control area) is essential in interpreting the amount of melanin pigment in disease states.	Increased darkness (brown to black) of skin surface over that considered normal
Hypopigmentation: decreased amount of melanin pigment in the epidermis and or superficial dermis	Decreased melanin pigment granules in the cytoplasm of keratinocytes or melanocytes. May be accompanied by pigmentary incontinence (presence of melanin pigment released from damaged epidermal melanocytes and basal cells, and phagocytized by dermal macrophages). Knowledge of the amount of melanin pigment in the normal skin (control area) is important in interpreting the amount of melanin pigment in disease states.	Lightening of skin surface from black or brown to grey, pink, or white.

DISEASE EXAMPLES

Numerous conditions as a post inflammatory change and affecting a variety of cutaneous locations (e.g., superficial spreading pyoderma and furunculosis)

Acanthosis nigricans

Primary (idiopathic)
Dachshunds, probably a genodermatosis

Secondary
A common reaction pattern beginning in but not limited to axillary areas and associated with friction or intertrigo, endocrine disease, hypersensitivity

Epitheliotropic lymphoma

Lupus erythematosus (Figure 3-21)

Post inflammatory or traumatic

Uveodermatologic syndrome (VKH-like disease)

Vitiligo

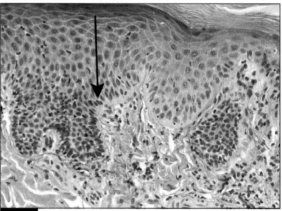

3-21 Depigmentation in skin of a dog with resolving case of lupus erythematosus. The arrow depicts the margin between pigmented and depigmented epidermis. Note epidermal pigment in macrophages in the dermis (pigmentary incontinence).

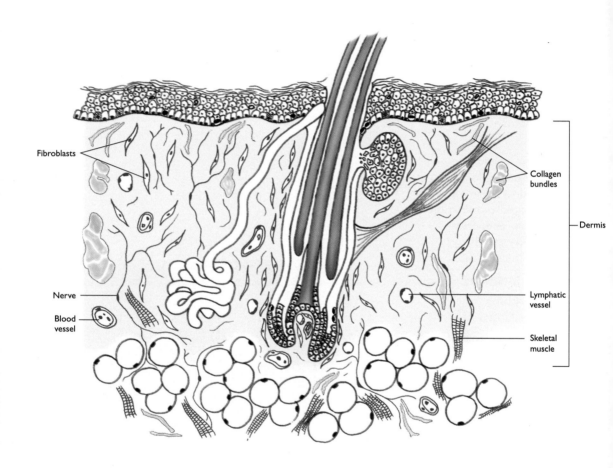

3-22 Dermis

Histopathologic Responses of the Dermis and Vessels to Injury

Dermis

✓The dermis consists of fibroblasts, collagen, and elastic fibers embedded in ground substance, and provides support for adnexa, vessels, and nerves (Figure 3-22). In the superficial dermis, the collagen fibers are smaller and finer than the larger collagen bundles in the deep dermis. The dermis also contains skeletal muscle fibers that extend from the underlying musculature and cause voluntary skin movement. A variety of other cells, most notably mast cells and lymphocytes are present in normal dermis. Plasma cells, macrophages and, more rarely, eosinophils and neutrophils, may be seen in normal dermis.

Vessels

✓Three vascular plexuses are in the skin: superficial, middle, and deep. The superficial plexus supplies the superficial portion of follicles (infundibulum) and epidermis. The middle plexus supplies middle portion of the hair follicles (isthmus), sebaceous glands, and arrector pili muscles. The deep plexus supplies the subcutis, inferior portion of hair follicles, and apocrine glands. Lymphatic capillaries originate in the superficial dermis, flow through the middle and deep dermis, and connect with a plexus in the panniculus. The lymphatic vessels then join, forming larger vessels that eventually reach peripheral lymph nodes.

Tables 3-5, 3-6, and 3-7 review the responses of the dermis and vessels to injury.

Table 3-5 Alterations in Growth or Development of the Dermis

RESPONSE OF THE DERMIS TO INJURY	HISTOLOGIC FEATURE	SALIENT CLINICAL FEATURE
Atrophy: reduction in thickness of the dermis	Decrease in amount of collagen fibrils and fibroblasts resulting in a reduced thickness of the dermis.	Palpably thin skin through which vessels may be seen The skin, especially in cats, may tear with minimal handling.
Fibroplasia: proliferation of fibroblasts and deposition of collagen fibrils	Increase in amount of collagen fibers and fibroblasts due to injury, including ulceration. In early lesions the fibroblasts are loosely woven and oriented perpendicular to proliferative capillaries (granulation tissue). In later stages collagen fibers are more compact and oriented in a laminar pattern to form a scar.	Localized area of thickening that gradually matures to form a scar. A scar may be palpable rather than visible if the haircoat covers the scar.
Collagen dysplasia: generally inherited abnormality of collagen	Microscopic features vary. In some types of collagen dysplasia syndromes, the dermis is normal in appearance, but in others the collagen bundles vary in size and shape and may be tangled with abnormal organizational pattern.	Skin that is hyperextensible and tears easily.

DISEASE EXAMPLES

Feline skin fragility syndrome
(Figure 3-23)

Hyperadrenocorticism

Topical or systemic steroid
therapy

Severe cachexia

Post incisional scar

Post inflammatory, associated
with severe furunculosis or
panniculitis

Post traumatic
chronic repetitive self
trauma, such as acral lick
dermatitis (Figure 3-24)

Ehlers-Danlos syndrome
(dermatosparaxis)

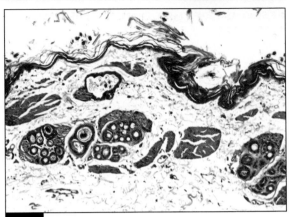

3-23 Severe dermal atrophy associated with feline skin fragility syndrome.

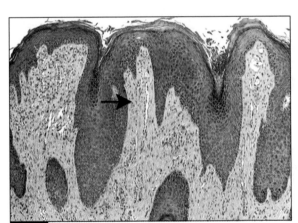

3-24 Fibroplasia in case of acral lick dermatitis. Note capillaries and collagen fibers oriented perpendicular to the epidermis (vertical streaking of collagen). The arrow is perpendicular to the vertically oriented collagen fibers. The epidermis is also thickened by compact hyperkeratosis and acanthosis.

continued

Table 3-5 Continued

RESPONSE OF THE DERMIS TO INJURY	HISTOLOGIC FEATURE	SALIENT CLINICAL FEATURE
Elastic tissue degeneration		
Solar elastosis: altered elastic tissue formed by fibroblasts chronically exposed to ultraviolet light	Solar elastosis consists of irregular, thickened, sometimes interwoven basophilic elastic fibers interspersed within dermal collagen fibers, typically in the superficial dermis.	Solar elastosis frequently occurs in conjunction with other changes (laminar fibrosis and acanthosis). Cumulatively these changes result in lichenified, indurated skin that may be erythematous.
	These altered elastic fibers stain basophilic (blue), which is in contrast to normal elastic fibers which stain eosinophilic (pink) and are indistinguishable from collagen fibers.	Solar keratoses and comedones may also be present. The comedones can progress to follicular cysts and furunculosis.
Elastosis of chronic moderate heat dermatitis: (erythema ab igne)—altered elastic tissue associated with exposure to moderate radiant or conductive heat	Chronic moderate heat dermatitis large keratinocyte nuclei, apoptotic basal cells and keratinocytes, brightly eosinophilic elastic tissue in dermis. The elastic fibers are more brightly eosinophilic than the collagen fibers.	Chronic moderate heat dermatitis mottled or reticulated erythema and alopecia

DISEASE EXAMPLES

Solar dermatitis
(Figures 3-25a and 3-25b)

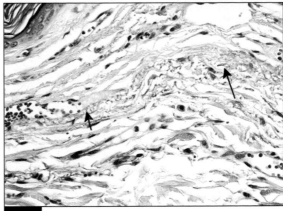

3-25a Solar dermatitis in dog. Note tangled bluish elastic tissue fibers in the dermis (arrows) in a section stained with hematoxylin and eosin. Damage from solar radiation causes this degenerative change in elastic fibers that serves as a marker for solar injury.

Chronic moderate heat
dermatitis
   (erythema ab igne)

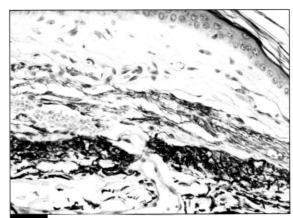

3-25b Solar elastosis in dermis of dog with solar dermatitis. Note the purplish-black elastic fibers in a section stained with Gomori's aldehyde fuchsin method.

Table 3-6 Degenerative or Depositional Disorders of Dermis

RESPONSE OF THE DERMIS TO INJURY	HISTOLOGIC FEATURE	SALIENT CLINICAL FEATURE
Collagen degeneration:* partially disrupted collagen fibers that may become mineralized.	Increased staining intensity or disruption of collagen fibers. Granular material represents degranulated eosinophils. Mineralization can occur.	Papules, plaques, or nodules. Often crusted, ulcerated, and inflamed.
Amyloid: protein that may be deposited in the dermis	Amyloid is an amorphous eosinophilic material. Special stains (congo red) help differentiate amyloid from collagen fibers.	Amyloid deposition is rare, and usually seen in association with cutaneous plasma cell tumor (clinically evident as nodule or tumor).
Mucin: protein bound to hyaluronic acid that may be deposited in the dermis. Hyaluronic acid has affinity for binding water.	Mucin is lightly basophilic amorphous material that may wash out of sections leaving collagen fibers widely separated from each other.	Mucin deposition results in puffy thickened skin. Mucin-filled vesicles may form. Mucin is thick stringy material that may ooze from needle pricked skin, from biopsy sites, or from ruptured vesicles.
Calcium: mineral deposits in the dermis either due to abnormality of dermal collagen, abnormal metabolism of calcium, phosphorous, and vitamin D, or for unknown reasons	Calcium deposits are irregular, granular, and basophilic and may be associated with a granulomatous and scarring inflammatory response.	Calcium deposition usually results in thick, firm, white to tan plaques with reddened borders. The plaques may be crusted.

* New information in cats suggests collagen fibers are partially disrupted but otherwise normal.

DISEASE EXAMPLES

Eosinophilic inflammation of a variety of causes

Feline eosinophilic granuloma
   (linear granuloma)
   (Figure 3-26)

Idiopathic eosinophilic granuloma of the dog

Mast cell tumor

Plasma cell tumor

Systemic disease

Mucin deposition may be seen in myxedema of hypothyroidism and in Shar pei mucinosis (Figure 3-27).

In the Shar pei breed, a small to moderate quantity of dermal mucin is considered normal. However, "pools" of mucin can accumulate in the dermis and, if extreme, may predispose the skin to easily tear and ooze mucinous fluid. These Shar pei dogs have dilated lymphatic channels and a reduced quantity of collagen.

Calcium deposition is usually seen in association with hyperadrenocorticism (calcinosis cutis) (Figure 3-28), and may develop focally in areas of persistent inflammation such as eosinophilic granuloma or an old granuloma.

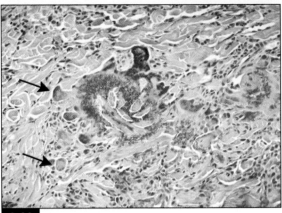

3-26 Feline eosinophilic granuloma. Brightly eosinophilic material consisting of collagen fibers and degranulated eosinophils is phagocytized by multinucleated giant cells (arrows).

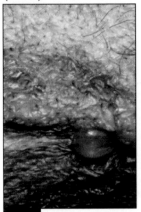
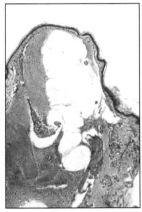

3-27 Clinical photograph of a vesicle from a Shar pei with vesicular mucinosis. Photomicrograph showing how the mucin has collected in large areas in the superficial to mid dermis causing the formation of dermal "pools" of mucin, seen clinically as vesicles.

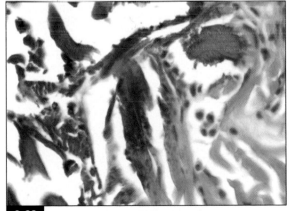

3-28 Dermal mineral deposition in poodle with calcinosis cutis associated with hyperadrenocorticism. The collagen becomes blue and granular as mineral accumulates.

Table 3-7 Inflammatory Disorders in the Dermis or Vessels

RESPONSE OF THE DERMIS TO INJURY	HISTOLOGIC FEATURE	SALIENT CLINICAL FEATURE
Perivascular inflammation	Perivascular accumulations of leukocytes bordering superficial, middle, and or deep vascular plexuses.	May be grossly visible as erythema and mild thickening (edema).
Vascular inflammation (vasculitis): Vasculitis is inflammation of a vessel wall in which the vessel is the primary target of injury	Necrosis of vessel wall limited to a few cells or patchy areas Infiltrates of leukocytes into the vessel wall Intramural or perivascular edema, hemorrhage, or fibrin exudation Vasculitis may have subsided by the time samples are collected. Evidence of pre-existing vasculitis (sometimes termed "footprints") in some syndromes includes adnexal atrophy.	Edema, erythematous macules, petechiae, ecchymoses, ischemic necrosis and infarction Classic vasculitis lesions are punctate ulcers on footpads and ears, scalloping of ear margins with sharp demarcation between normal and abnormal tissue. Ulceration of distal extremities (pinnae, toes, and tail tip) with or without sloughing of tissue.
Interface Inflammation: cell rich cell poor	Inflammation at the epidermal dermal interface with vacuolar or apoptotic basal cell degeneration. The inflammation may be mild (cell poor) or abundant (cell rich).	Interface lesions cause weakening of the epidermal dermal attachment and may result in vesicles, erosions, or ulcers. Hypopigmentation also may be a feature. The skin may be thickened with reduction or loss of the normal epidermal architecture (reduced skin creases).

DISEASE EXAMPLES

Common finding in many inflammatory conditions. As a sole feature it is not highly diagnostic (Figure 3-29).

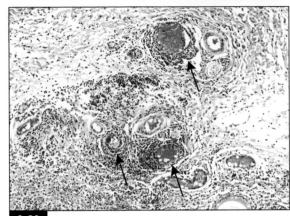

3-29 Perivascular inflammation. Inflammatory cells are concentrated around dermal blood vessels (arrows).

Dermatomyositis

Idiopathic vasculitis (Figure 3-30)

Immune complex deposition (systemic lupus erythematosus)

Infection with endotheliotropic organisms (*Rickettsia rickettsii*)

Septicemia (disseminated associated with infection)

Venomous spider bite

Vaccine-induced ischemic dermatitis

Vasculitis in Jack Russell terriers, Greyhounds, German shepherds, etc.

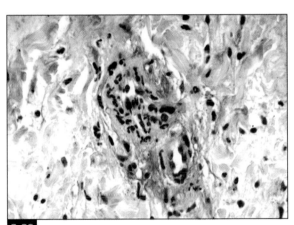

3-30 Vasculitis in a dog. Neutrophilic inflammation is directed toward the vessel wall, and there is mucinous fluid in and around the vessel wall. The cause of the vasculitis was undetermined.

Cell poor
   Dermatomyositis
   Vaccine-induced ischemic
     dermatitis
   Vasculitis in Jack Russell
     terriers

   Vitiligo
     Minor interface
     inflammation and
     absence of basal cell
     degeneration

Cell poor or cell rich
   Lupus erythematosus

continued

Table 3-7 Continued

RESPONSE OF THE DERMIS TO INJURY	HISTOLOGIC FEATURE	SALIENT CLINICAL FEATURE
Lichenoid Inflammation: **Controversial term** **1. Presence of basal cell degeneration** versus **2. Absence of basal cell degeneration**	Dense band-like accumulation of inflammatory cells just below the epidermis. Some pathologists use the term "lichenoid inflammation" synonymously with cell rich interface dermatitis to refer to diseases with basal cell degeneration. In contrast, other pathologists use this term to refer to dense inflammation below the epidermis, without basal cell degeneration.	Tissue may be thickened with flattening or loss of the normal epidermal architecture (reduced skin creases) Hypopigmentation also may be a feature.
Nodular to diffuse granulo-matous inflammatory lesions (infectious):	Nodular accumulation of mixed inflammatory cells including macrophages. Macrophages may contain infectious agents in the cytoplasm. Organisms may also be free in the dermal stroma.	Nodular or plaque-like lesions that may be alopecic, ulcerated, or draining

DISEASE EXAMPLES

With basal cell degeneration
   Discoid lupus erythematosus

Without basal cell degeneration
   Epitheliotropic lymphoma
   lymphocytes are present in
   the epidermis and may mimic
   inflammation

   Idiopathic lichenoid dermatitis

   Intertriginous pyoderma

   Mucocutaneous pyoderma
   (Figure 3-31)

   Lichenoid psoriasiform
   dermatitis of Springer spaniel

   Uveodermatologic syndrome
   (Vogt-Koyanagi-Harada-like
   syndrome)

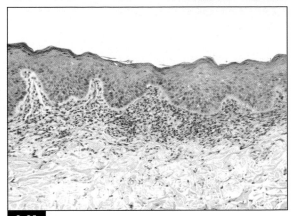

3-31 Inflammation in superficial dermis of dog with mucocutaneous pyoderma. Note dense band of inflammation that is parallel to the epidermis. The epidermal dermal interface is largely spared.

Granulomatous dermatitis
associated with infectious agents
such as bacteria, fungi, algae, or
protozoa (Figure 3-32)

Examples:

   Blastomycosis

   Leishmaniasis

   Nocardiosis

   Protothecosis

   Sporotrichosis

   Yersinia pestis and cowpox
   in cats (rare and geographic)

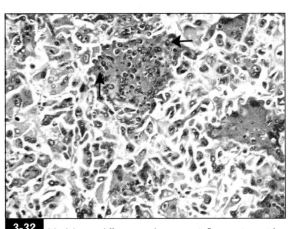

3-32 Nodular to diffuse granulomatous inflammation with dematiaceous (pigmented) fungal organisms (fungi are brown in unstained or H&E sections) (arrows).

continued

Table 3-7 Continued

RESPONSE OF THE DERMIS TO INJURY	HISTOLOGIC FEATURE	SALIENT CLINICAL FEATURE
Nodular to diffuse granulomatous inflammatory lesions (noninfectious):	Nodular accumulation of mixed inflammatory cells including macrophages. No organisms are identified. Diagnosis of sterile inflammation is made only after ruling out infectious agents (e.g., performing special stains on histologic sections, cultures of deep tissue biopsy samples, and lack of response to appropriate therapy). Examples of additional tests that can be performed for some infectious agents include immunohistochemistry, immunofluorescence, polymerase chain reaction, and in situ hybridization.	Nodular or plaque-like lesions that may be alopecic, ulcerated, or draining.
Nodular to diffuse inflammation with prominent eosinophils:	Nodular to diffuse inflammation with large numbers of eosinophils. Some lesions have degranulated eosinophils bordering collagen fibers (flame figures). Eosinophilic plaques frequently have mucinous fluid in the epidermis and follicular epithelium.	Irregular or linear plaques, patches, nodules, or ulcers
Nodular to diffuse inflammation with prominent plasma cells:	Nodular to diffuse inflammation with prominent plasma cells. Some plasma cells may contain abundant cytoplasmic immunoglobulin. Ulcerated lesions may contain neutrophils.	If involving the footpads of cats, lesions are frequently swollen, puffy or spongy, and may ulcerate.

DISEASE EXAMPLES

Juvenile sterile granulomatous dermatitis and lymphadenitis (Figure 3-33)

Sterile granuloma and sterile pyogranuloma syndrome

Xanthogranuloma

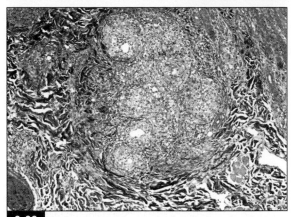

3-33 Nodular to diffuse granulomatous inflammation with no organisms in puppy with juvenile, sterile, granulomatous dermatitis and lymphadenitis. The inflammation has a periappendageal orientation.

Canine eosinophilic granuloma

Dermatitis of a variety of causes with prominent eosinophils (e.g., parasite bites)

Feline eosinophilic granuloma (linear granuloma)

Feline eosinophilic plaque (Figure 3-34)

Feline eosinophilic ulcer (indolent ulcer, rodent ulcer)

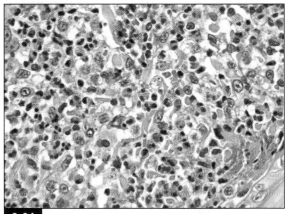

3-34 Nodular to diffuse inflammation with prominent eosinophils in the dermis of a cat with an eosinophilic plaque.

Feline plasma cell pododermatitis (Figure 3-35)

Persistent antigenic stimulation from chronic infection or trauma

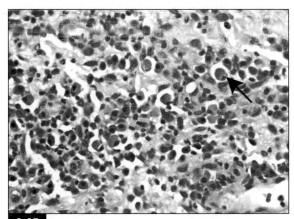

3-35 Nodular to diffuse inflammation with prominent plasma cells in the dermis of a footpad of a cat with plasma cell pododermatitis. Note the droplets of cytoplasmic immunoglobulin in the plasma cells (Russell bodies)(arrow).

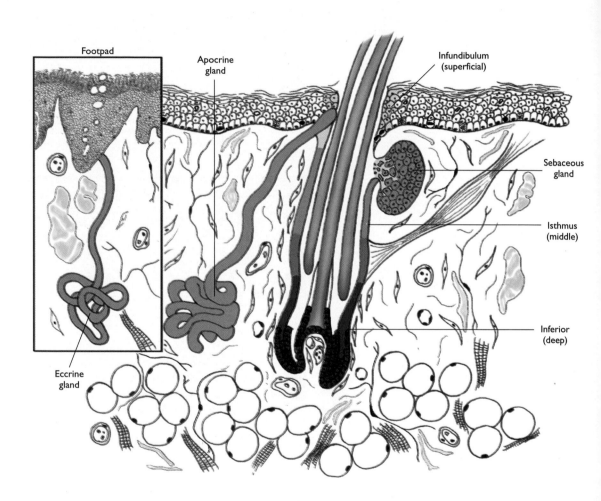

3-36 Adnexa

Histopathologic Responses of the Adnexa to Injury

Adnexa

Hair Follicles

✓ The hair cycle of dogs and cats has a growth phase (anagen), a transitional phase (catagen), and a resting phase (telogen). Dogs and cats have compound follicles that consist of large primary follicles and smaller secondary follicles. These follicular units are oriented with primary hairs toward the front of the body and secondary hairs toward the rear. Biopsy sampling methods to facilitate examination of hair follicle morphology are described in Section 1 (see Figure 1-1). There are a variety of ways of subdividing hair follicles anatomically. The traditional way divides the follicle into superficial, middle, and deep portions (Figure 3-36). The superficial portion is termed the infundibulum, and it extends from the follicular opening to the level of the entrance of the sebaceous gland duct. The epithelium of the infundibulum is identical to and continuous with the epidermis. The middle portion is termed the isthmus, and it extends from the level of the entrance of the sebaceous duct to the attachment of the arrector pili muscle. The epithelium of the isthmus keratinizes without formation of a keratohyalin layer, through a process termed tricholemmal keratinization, to an amorphous layer of pale staining keratin. The deep portion is termed the inferior portion. It is the portion below the attachment of the arrector pili muscle and includes the hair bulb. The hair bulb is composed of epithelial cells, also called hair matrix cells, which border the papilla.

Sweat Glands

✓ Apocrine glands are also termed epitrichial glands because the ducts open in the superficial portion of the hair follicle (Figure 3-36). In contrast, the ducts of the eccrine glands, termed atrichial glands, open directly onto the surface of the skin. Apocrine glands are tubular or saccular coiled glands lined by secretory epithelium. The apocrine glands are located in haired areas of skin, in the external ear canal and eyelids, and comprise the anal sac glands. Eccrine glands are tubular glands lined by cuboidal epithelium, and are confined mainly to footpads of dogs and cats.

Sebaceous Glands

✓ Sebaceous glands are simple, branched, or compound alveolar glands (Figure 3-36). The ducts open into hair follicles and, at some mucocutaneous junctions, directly on the skin surface. Sebaceous gland secretion is holocrine, that is the entire cell with secretory contents is shed into the duct. Perianal (circumanal, hepatoid) glands are modified sebaceous glands that lack patent ducts. Perianal glands occur most commonly near the anus, but are also present in skin near the prepuce, tail, flank, and groin. Perianal glands are composed of lobules of cells resembling hepatocytes, thus the synonym hepatoid glands.

Tables 3-8 and 3-9 review the responses of the adnexa to injury.

Table 3-8 Alterations in Growth or Development of Adnexa

RESPONSE OF THE ADNEXA TO INJURY	HISTOLOGIC FEATURE	SALIENT CLINICAL FEATURE
Follicular hyperkeratosis: histologic term for increased quantity of stratum corneum within follicle lumens.	Superficial to mid portion of follicle dilated and distended with cornified cells.	Fronds of stratum corneum may be adherent to hair shafts as they exit from follicular openings (follicular casts). Follicular opening may appear more prominent. Comedones ("blackheads") may be present.
Acanthosis: thickening of the spinous layer of the follicular infundibular epithelium.	The follicular infundibular epithelium is thickened by cells of the stratum spinosum.	Mild acanthosis may result in no gross change. Usually associated with hyperplasia of the epidermis resulting in lichenification.
Hypertrophy: Increase in size of a structure	Typically refers to increase in size of a follicle and usually involves an associated increase in the size of sebaceous and apocrine glands.	May be no gross change with this lesion alone. Usually associated with increased thickness of epidermis and dermal collagen resulting in lichenification or plaque-like area of thickening.

DISEASE EXAMPLES

Actinic comedones

Chin acne

Endocrine dermatoses

Primary seborrhea

Schnauzer comedo syndrome
(Figure 3-37)

Sebaceous adenitis

Surface trauma (e.g., callus)

Vitamin A responsive
dermatosis

Conditions that cause
acanthosis of the epidermis such
as chronic inflammation and
surface trauma also result in
follicular acanthosis.

Acral lick dermatitis

Chronic allergic dermatitis

Acral lick dermatitis

Chronic allergic dermatitis

Follicular hamartoma
(Figure 3-38)

3-37 Follicular hyperkeratosis and dilation in Schnauzer with Schnauzer comedo syndrome. A plug of keratin at the follicular orifice may be noted clinically as a black spot.

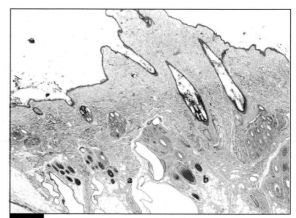

3-38 Follicular hypertrophy in a dog with follicular hamartoma. Note the large hair bulbs (b) in comparison to those in the adjacent normal skin (a).

continued

Table 3-8 Continued

RESPONSE OF THE ADNEXA TO INJURY	HISTOLOGIC FEATURE	SALIENT CLINICAL FEATURE
Atrophy: reduction in the size of a structure. Technically, the term atrophy is defined as follicles smaller than expected for the physiologic stage of the hair cycle. However, atrophy also may be used when referring to a greater than expected number of follicles in non anagen stages of the hair cycle.	Follicles are inactive and small. There frequently are increased numbers of inactive follicles over that expected for the stage of the hair cycle.	Alopecia (no hair) Hypotrichosis (reduced amount of hair, e.g., thinning of the haircoat).
Abnormalities associated with hair cycle stages (other than atrophy as defined above):		
Telogen: resting stage	**Telogen defluxion (effluvium)** Telogenization—all follicles enter into telogen stage synchronously. In recovery, follicles are in the growth stage.	**Telogen defluxion (effluvium)** Usually seen clinically 1 to 3 months after stressful event when a new hair cycle begins, and new hair shafts grow pushing old hair shafts out (causes sudden shedding of older hair shafts). If clipped during the telogen stage, hair will fail to regrow.
Catagen: transition stage	**Catagen or telogen arrest** Follicles in catagen or telogen stage of the cycle with prominent tricholemmal keratin (flame follicles).	**Catagen or telogen arrest** Hair fails to regrow after close clipping or shaving.
Anagen: growth stage	**Anagen defluxion** Degeneration of hair matrix cells of hair follicle	**Anagen defluxion** Sudden loss of anagen hair shafts.
Dysplasia: abnormal development of adnexa, especially follicles.	Microscopic lesions vary with type Pigment clumps in hair matrix cells and hair shafts Small (petite, miniaturized) diminutive hair bulbs Irregular distorted hair follicles	**Alopecia or hypotrichosis**

DISEASE EXAMPLES

Alopecia X (alopecia of plush coated breeds)

Endocrine disease

Feline paraneoplastic alopecia

Feline pinnal alopecia

Idiopathic (cyclic) flank alopecia syndrome

Ischemic lesions (vascular)
  dermatomyositis
  vaccine-induced ischemic
    dermatitis (Figure 3-39)
  vasculitis in Jack Russell terriers

Secondary to follicular dysplasia or sebaceous adenitis

Traction alopecia

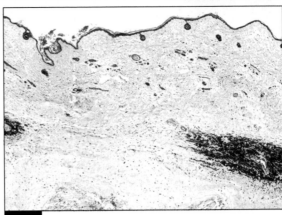

3-39 Follicular atrophy in dog with post rabies vaccination dermatitis. Note lymphocytic inflammation in the deep dermis and panniculus.

Telogen defluxion (effluvium) Stressful event (e.g., fever, pregnancy, severe illness)

Catagen or telogen arrest Post clipping alopecia usually of long or plush coated breeds of dogs

Anagen defluxion (uncommon) Antimitotic drug therapy or other event, usually seen in dogs with continuously growing haircoats such as poodles.

Alopecia of curly coated retriever

Alopecia of Portuguese water dog

Black hair follicle dysplasia

Color dilution alopecia (follicular dysplasia) (Figure 3-40)

Pattern alopecia

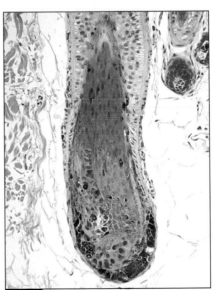

3-40 Melanin pigment clumps displacing hair matrix cells in a blue Doberman pinscher with color dilution alopecia (follicular dysplasia).

Luminal folliculitis

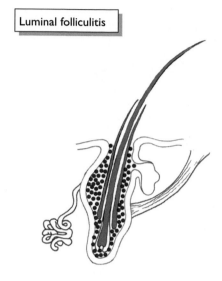

Mural folliculitis

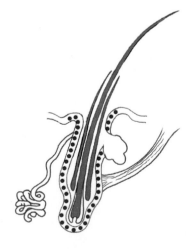

3-41a Types of folliculitis (luminal, mural, and mural subtypes). KEY POINT: Reaching a specific diagnosis can be facilitated by the identification of the type of folliculitis. For instance, luminal folliculitis usually indicates infectious folliculitis. (see Table 3-9.)

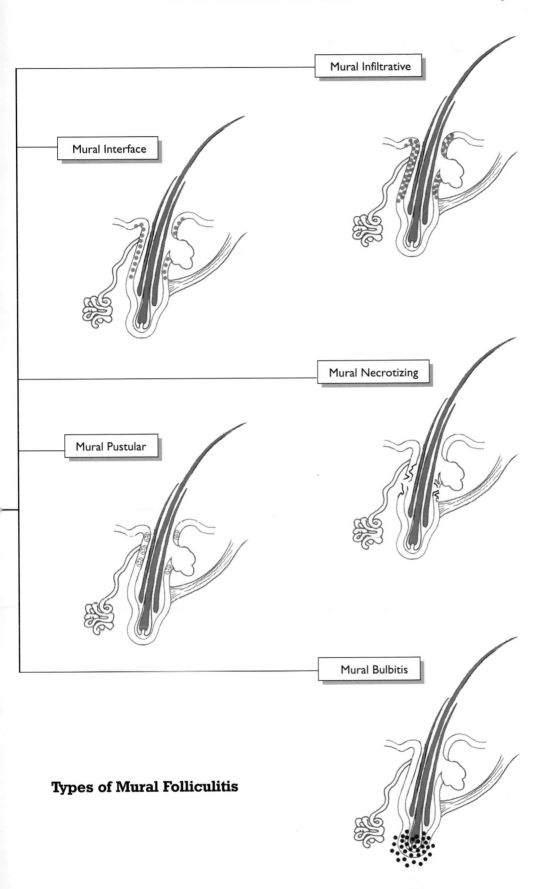

Types of Mural Folliculitis

Mural Infiltrative

Mural Interface

Mural Necrotizing

Mural Pustular

Mural Bulbitis

Table 3-9 Inflammatory Disorders of the Adnexa

RESPONSE OF THE ADNEXA TO INJURY	HISTOLOGIC FEATURE	SALIENT CLINICAL FEATURE
Perifolliculitis: inflammation surrounding a hair follicle, but not involving the follicle.	Leukocytes occur around perifollicular vessels, hair follicles, and if severe, the adjacent glands.	Early gross lesion is a papule.

Folliculitis:* inflammation of a hair follicle (see Figure 3-41a)
 	
Further defined by follicular anatomy:
 	
Superficial (infundibulum = opening to entrance of sebaceous duct)
 	
Middle (isthmus = entrance of sebaceous duct to attachment of arrector pili muscle)
 	
Inferior (below isthmus and including the bulb)
 	
Luminal (inflammation of the wall and lumen of a hair follicle)
  Infectious
  Sterile
 	
Mural folliculitis (refers more stringently to inflammation of the follicular wall but not lumen)
  Interface
  Infiltrative
  Pustular
  Necrotizing
  Bulbitis (hair bulb)
	Luminal Inflammation of the follicular lumen at any level.
	
	Interface mural The outer (basilar) epithelium and superficial to mid portion are most commonly affected.
	
	Infiltrative The basilar and spinous epithelium of the superficial to mid portion are most commonly affected.
	
	Necrotizing The entire wall of the middle to inferior portion are most commonly affected.
	
	Pustular The subcorneal epithelium of the superficial to mid portion is most commonly affected.
	
	Bulbitis Inflammation of the hair bulb and peribulbar dermis
		Luminal Fully developed luminal folliculitis may be visible as follicular papules and pustules.
		
		Small white plug of debris (intrafollicular) may be seen.
		
		Mural Mural folliculitis appears as alopecia or hypotrichosis.

DISEASE EXAMPLES

Early lesion that usually precedes development of folliculitis of a variety of causes (e.g., staphylococcal folliculitis).

Can be seen with sterile lesions such as multinodular periadnexal dermatitis.

Luminal (Figure 3-41b)
Infectious
(bacterial dermatophyte, parasitic)

Sterile
eosinophilic folliculitis

Interface mural (Figure 3-41c)
Demodicosis

Dermatomyositis

Erythema multiforme

Graft versus host disease

Lupus erythematosus

Vaccine-induced ischemic dermatitis

Vasculitis in Jack Russell terriers

Infiltrative (Figure 3-41d)
Epitheliotropic lymphoma

Feline mural folliculitis
Unknown

Pustular (Figure 3-41e)
Pemphigus follaceus and erythematosus

Necrotizing (Figure 3-41f)
Drug eruptions

Eosinophilic folliculitis and furunculosis

Feline mosquito bite hypersensitivity

Herpesvirus infection

Bulbitis (Figure 3-41g)
Alopecia areata

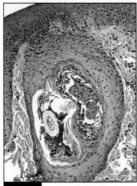

3-41b Luminal folliculitis in a case of superficial bacterial pyoderma. Note the inflammatory cells located predominantly within the lumen of the follicle.

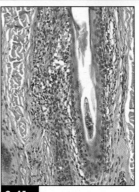

3-41c Interface mural folliculitis in a case of canine demodicosis. Note *Demodex canis* within the follicular lumen and inflammation along the epithelial dermal interface of the hair follicle.

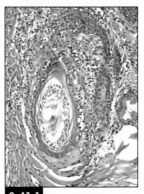

3-41d Infiltrative mural folliculitis in a dog. Note the architecture of the follicular wall is disrupted by inflammatory cells and spongiosis. The follicular lumen contains no inflammatory cells.

3-41e Pustular folliculitis in a dog with pemphigus foliaceus. Note accumulation of inflammatory cells and acantholytic keratinocytes within the follicular wall.

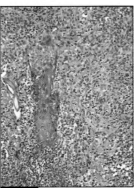

3-41f Necrotizing folliculitis in a case of feline herpesvirus dermatitis. Herpesvirus is epitheliotropic. It causes necrosis of the epidermis and sebaceous gland epithelium as well as necrosis of the follicular wall. Necrotizing mural folliculitis is only a component of the lesions in this disease.

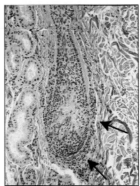

3-41g Inflammation targeting the anagen hair bulbs in a dog with alopecia areata. The lymphocytes around the hair bulb (arrows) have been termed, "a swarm of bees".

continued

Table 3-9 Continued

RESPONSE OF THE ADNEXA TO INJURY	HISTOLOGIC FEATURE	SALIENT CLINICAL FEATURE
Furunculosis: inflammation of a follicular wall associated with follicular rupture.	Inflammation and disruption of the follicular wall with release of hair and cornified cells into the surrounding dermis. Nodular inflammation with fistulae may develop. There may be extension of the inflammation into the panniculus.	Follicular papules, pustules, hemorrhagic bullae, crusts, ulcers, painful nodules, fistulae, alopecia. Small white plug of debris (intrafollicular) may be seen.
Sebaceous adenitis: inflammation of sebaceous glands. May be seen as extension of folliculitis or other inflammatory process near follicles or as an independent (primary) condition.	Primary sebaceous adenitis—Inflammation directed toward and specifically involving sebaceous glands and ducts; may be accompanied by hyperkeratosis and follicular atrophy. Secondary sebaceous adenitis—Inflammation that extends from adjacent tissue to involve sebaceous glands.	Scaling and follicular casts (corneal cells adherent to hair shafts) are early lesions, and then alopecia develops. Secondary deep pyoderma may be seen, especially in long haired breeds. Blepharitis is also a feature in conjunction with other changes described above.
Hidradenitis: inflammation of apocrine glands usually seen in association with suppurative or granulomatous dermatitis.	Apocrine glands are dilated with secretory product and inflammatory cells, and are bordered by inflammatory cells.	No gross lesion by itself, but may accompany draining tracts or nodular lesions.

DISEASE EXAMPLES

Progression of infectious folliculitis or sterile folliculitis, follicular cysts, or drug reactions.

Infectious folliculitis is most frequently caused by *Demodex* mites, staphylococci (Figure 3-42), dermatophytes, but there are many other causes (*e.g., Pelodera* sp.).

Primary idiopathic sebaceous adenitis (Figure 3-43)

Secondary sebaceous gland inflammation as extension from folliculitis or inflammation near follicles

Demodicosis

Early canine reactive histiocytosis

Exfoliative cutaneous lupus erythematosus (lupoid dermatosis of the German short-hair pointer)

Idiopathic multinodular periadnexal dermatitis

Leishmaniasis

Staphylococcal folliculitis

Uveodermatologic syndrome

Often associated with obstructive follicular inflammatory processes such as chronic pyoderma.

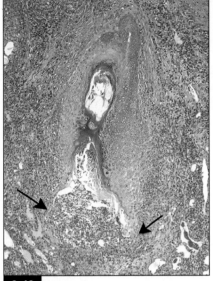

3-42 Furunculosis in case of staphylococcal associated folliculitis. Note the disrupted follicular wall (arrows).

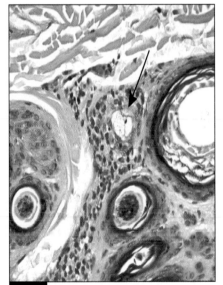

3-43 Active sebaceous adenitis in a dog. Note inflammation targeting the sebaceous gland (arrow).

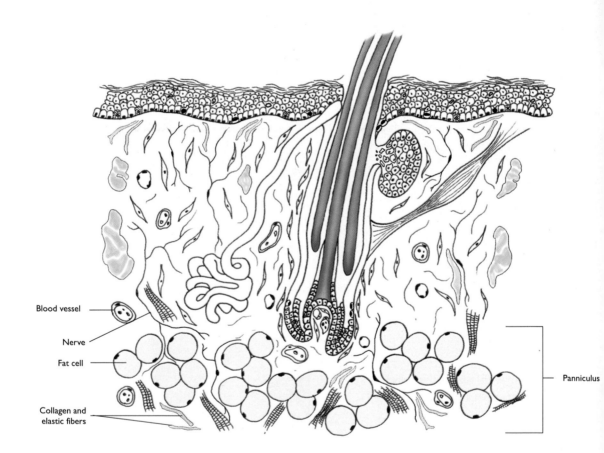

Blood vessel

Nerve

Fat cell

Collagen and
elastic fibers

Panniculus

3-44 Panniculus

Histopathologic Responses of the Panniculus (Subcutis) to Injury

✓The panniculus is composed of lobules of adipose tissue subdivided by collagenous septa containing vessels, nerves, and elastic fibers (Figure 3-44). The panniculus attaches the dermis to underlying tissue, including muscle or bone. In human medicine much emphasis is placed on the pattern of inflammation (*e.g.*, predominantly affecting lobules versus septa) because this has diagnostic significance. In veterinary medicine, the lobular versus septal pattern does not have diagnostic significance; therefore, the emphasis in diagnosis is placed on the predominate type of inflammatory cell in the lesions. With the exception of injection site reactions, pathologic processes targeting the panniculus are relatively uncommon in dogs and cats. Most cases of panniculitis result from an extension of dermal inflammation into the subcutis.

Table 3-10 reviews inflammatory lesions of the panniculus.

Table 3-10 Inflammation of the Panniculus (Panniculitis)

RESPONSE TO INJURY	HISTOLOGIC FEATURE	SALIENT CLINICAL FEATURE
Disorders of the panniculus: Panniculitis is inflammation targeting the subcutaneous adipose tissue. Panniculitis may be primary or secondary (e.g., extension from the dermis).	Inflammation in fat lobules or interlobular septa or both. Predominantly neutrophilic Predominantly lymphocytic Predominantly pyogranulomatous With microorganisms Without microorganisms	Nodular lesions that may ulcerate and drain oily or blood tinged material. Lesions may be single (e.g., abscess, vaccine reaction) or multifocal (e.g., idiopathic sterile nodular panniculitis, pancreatic panniculitis).

DISEASE EXAMPLES

Predominantly neutrophilic
Abscess

Foreign body reaction

Pancreatic panniculitis associated with pancreatitis or pancreatic carcinoma (Figure 3-45)

Feline nutritional pansteatitis

Predominantly lymphocytic
Lupus panniculitis

Vaccine-induced dermatitis and/or panniculitis

Granulomatous to pyogranulomatous with infectious agents

Atypical mycobacterial infection (Figure 3-46)

Granulomatous to pyogranulomatous without infectious agents

Idiopathic sterile nodular panniculitis

Metatarsal fistulation of German shepherds (sterile pedal panniculitis)

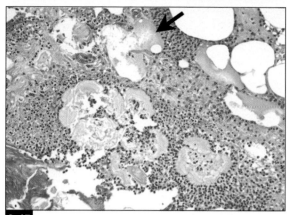

3-45 Pancreatic panniculitis. Suppurative exudate borders and separates adipose tissue of the panniculus. Note fat necrosis in areas of inflammation and represented by the granular material (arrow).

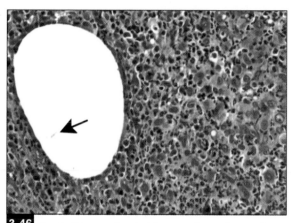

3-46 Panniculitis associated with atypical mycobacterial infection. Note the faintly staining organisms within the fat vacuole (arrow).

Section 4

Tumors and Tumor-like
Masses of the Skin

Introduction

This chapter does not provide a thorough discussion of tumors, but a brief listing of salient features of primary skin tumors and tumor-like lesions, as well as information pertaining to diagnosis and clinical behavior. Methods of biopsy sampling or excision are discussed in Section 1. Table 4-1 lists the general features of benign versus malignant tumors of the skin, and Table 4-2 lists tools that aid in the diagnosis of tumors of the skin. Tables 4-3, 4-4, 4-5 and 4-7 provide a brief listing of primary cutaneous tumors arising in the different levels within the skin (epidermis, adnexa, dermis, and subcutis). Table 4-6 illustrates grading systems for mast cell tumors, and Tables 4-8 through 4-10 list cysts, hamartomas, and other nodular tumor-like lesions. It should be realized that the names of tumors may change either because more is learned about the cell of origin of a tumor or because classification systems are refined. Therefore, there may be some variation in tumor terminology among pathologists. If there is any confusion about the name or type of tumor discussed in a pathology report, the pathologist providing the report should be consulted for clarification. The authors have attempted to use the most accepted terminology, and to provide synonyms when appropriate. Synonyms were omitted in rare instances when synonym terminology was considered controversial, and prognosis and therapy for those tumors were different.

Especially when evaluating Tables 4-3, 4-4, 4-5 and 4-7, it is helpful to remember the embryological derivations of cells within the skin, because frequently tumors arising from the same cell line exhibit similar behavior with regard to local invasion and metastasis. As such, the ectodermal origin of the epidermal layer and hair follicles, the mesodermal cell origin of much of the dermal layer, and the neuroectodermal origin of melanocytes can help us to explain their differences in behavior once neoplastically transformed. This is perhaps most true for the mesodermal tumors, in which metastatic occurrence is relatively low, yet local recurrence rates can be high, particularly with fibrosarcomas. Various **histologic** grading systems have been developed to help correlate histologic features of specific tumors with tumor behavior and thus with prognosis (Table 4-6). In addition, systems for staging neoplasms **clinically** have been developed to help standardize treatment regimes and evaluations of response to therapy. Complete clinical staging includes thoracic radiographs, abdominal radiographs and/or ultrasonography, complete blood count, biochemical profile, urinalysis, lymph node aspirate or biopsy, and a bone marrow aspirate. Extra care should be taken when using invasive testing in animals with suspected bone marrow infiltrative disease, as these animals may have thrombocytopenia, which may lead to increased bleeding time. Enlargement of regional lymph nodes is an indication for lymph node biopsy. A commonly used staging system was developed by the World Health Organization (Table 4-2). This system uses three categories to define tumor stages: 1) size of the primary tumor {P}, 2) presence or absence of lymph node involvement {N}, and 3) presence or absence of metastasis {M}.

Tumors of the Skin

A definition of tumor that has withstood the test of time is "an abnormal mass of tissue, the growth of which exceeds and is uncoordinated with that of normal tissue, and persists in the same excessive manner after cessation of the stimuli which evoked the change" (Willis, 1948). The skin is a common site of tumor development in dogs and cats. Most cutaneous tumors originate in the skin and are termed, "primary skin tumors". However, tumors may arise in other tissues and secondarily involve the skin. Examples include invasion of the adjacent skin by mammary carcinomas, and metastasis to the skin of more distant tumors such as canine hemangiosarcoma and feline bronchogenic carcinoma. Clinical appearances may vary from a discrete nodular mass (Figure 4-1) to a poorly delineated and infiltrative mass, to lesions that have the clinical appearance of inflammation (Figure 4-2) (such as diffuse or regional erythematous and exfoliative or exudative lesions). Thus the clinical evaluation of cutaneous tumors often presents a diagnostic challenge. The most common cutaneous tumors in cats are basal cell tumor of epidermis and adnexa, squamous cell carcinoma, fibrosarcoma, and mast cell tumor. The most common cutaneous tumors in dogs are hepatoid gland adenoma, sebaceous gland adenoma, cutaneous histiocytoma, mast cell tumor, and lipoma.

Cutaneous tumors originate in different components of the skin. Most tumors of epidermal and adnexal origin (ectodermal) are benign. Notable exceptions are tumors of the apocrine glands (sweat glands and anal sac glands), and squamous cell carcinomas. In contrast, tumors of leukocytes, fibrous tissue, muscle, and vessels (mesodermal), are more often malignant exhibiting infiltrative growth and sometimes metastasis. Melanocytic tumors (neuroectodermal) with a similar histologic appearance may exhibit different biological behavior, depending in part on their anatomic location. For example, those arising in the haired skin are more often benign (melanocytoma, Figure 4-3), whereas those arising in the oral cavity or subungual areas are more often malignant. A high mitotic index is a further major feature for judging malignancy.

4-1 **Sebaceous adenoma,** a discrete, irregular mass protuding from the skin surface. These tumors may be partly covered by waxy secretion. They are frequently multiple and occur in old dogs.

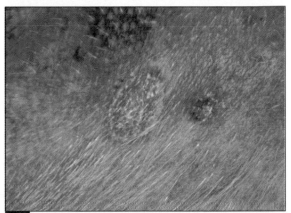

4-2 **Epitheliotropic lymphoma** manifesting as epidermal collarettes and erythema. Epitheliotropic lymphoma has a variable clinical presentation including: 1) erythematous scaly skin, 2) mucocutaneous ulceration and depigmentation, 3) single or multiple raised cutaneous masses, 4) ulcerative mucosa. Note clinical similarity to Figure 2-8a.

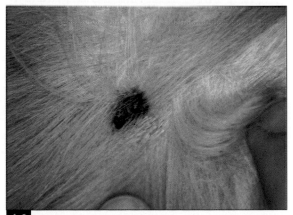

4-3 **Melanocytoma,** a pigmented solitary nodule from the lateral thigh of a dog. The tumor was moveable in the skin.

4-4 **Fibrosarcoma,** an irregular multifocally ulcerated mass arising in the site of amputation of a front leg in a cat.

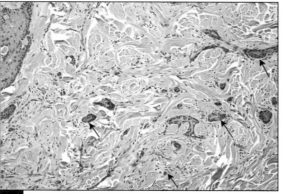

4-5 **Squamous cell carcinoma** in the skin of a dog. Note small nests of tumor cells (arrows) in the dermal collagen (invasion).

Euthanasia may be performed due to metastasis, but is also commonly due to local invasion and tumor recurrence (Figure 4-4). Therefore, evaluation of tumor margins for completeness of excision, and knowledge of tumor behavior are important in predicting prognosis. Histopathologic evaluation of tumors is not only regarded as a valuable diagnostic tool, but allows the accurate assessment of surgical margins (Figure 4-5), and lymphatic and blood vessels can be evaluated for metastatic spread (tumor cells) (Figure 4-6). These features can not be evaluated by examination of impression smears or aspirates of tumors.

In some poorly differentiated tumors, the cell of origin cannot always be determined in the standard hematoxylin and eosin stained sections (Figure 4-7). Numerous histochemical staining procedures (termed, "special stains") have been developed in order to help identify the cell of origin. The majority of these histochemical stains can be performed on formalin fixed paraffin embedded tissues, and are listed in section 5. In addition, immunologic (immunohisto-chemical) staining procedures that use the specificity of one or a group of antibodies directed against cell surface or cytoplasmic proteins have also been developed to help determine cell of origin of poorly differentiated tumors (see section 5). Some of the immunohistochemical staining procedures can be performed on formalin fixed paraffin embedded tissues, but fresh or snap frozen specimens are usually preferred.

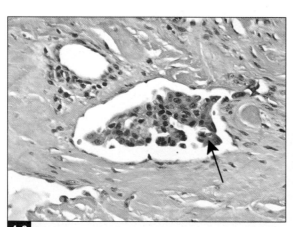

4-6 **Squamous cell carcinoma** in the skin of a dog. Note tumor cells are within a thin walled lymphatic vessel (arrow).

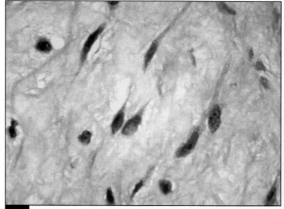

4-7 **Melanoma, spindle cell variant.** The spindle cells lack melanin granules. With H&E sections, these cells are difficult to distinguish from other spindle cell skin tumors such as fibrosarcoma or spindle cell squamous cell carcinoma, which typically have different biologic behavior.

Table 4-1
General Features of Benign and Malignant Tumors of the Skin

	BENIGN TUMOR	MALIGNANT
Clinical appearance*	Discrete, circumscribed, sometimes encapsulated, moveable nodular mass	Poorly defined, invasive nodular mass fixed to surrounding tissues
Histopathology	Well differentiated (resemble tissue of origin) Often have a low mitotic index	Locally infiltrative Poorly differentiated (anaplastic = vary in cell size and shape; large, vesicular nucleus; large nucleoli of increased number) Often have high mitotic index
Behavior	Remain localized (do not spread)	Principal criterion of malignancy is ability to metastasize Local invasion may result in incomplete removal and recurrence

*Tumors may mimic inflammation, immune mediated disease, or ulcers and non-healing wounds. Mimickers of inflammation: erythematous or exfoliative lesion (apocrine gland carcinoma, epitheliotropic lymphoma) and localized areas of swelling that wax and wane (some mast cell tumors). Mimicker of autoimmune disease: erythematous and/or depigmenting lesion that may affect oral cavity, mucocutaneous junctions, and nose or footpads (epitheliotropic lymphoma). Mimicker of ulcer or non-healing wound (squamous cell carcinoma, other tumors)

Table 4-2
Tools that Aid in the Diagnosis and Prognosis of Tumors

TOOL	MEASURE	EXAMPLE
Physical exam Radiographs, Ultrasound **Clinical staging systems** help standardize treatment and evaluation of response to therapy. Example is World Health Organization TNM staging system (see column 3).	Size of primary tumor	T_0 No evidence of neoplasia
		T_1 Tumor <1 cm, not invasive
		T_2 Tumor 1-3 cm, locally invasive
		T_3 Tumor >3 cm, or evidence of ulceration or local invasion
	Presence or absence of lymph node involvement	N_0 No evidence of lymph node involvement
		N_1 Node firm, enlarged
		N_2 Node firm, enlarged, and fixed to surrounding tissues
		N_3 Nodal involvement beyond first station
	Presence or absence of metastasis	M_0 No evidence of metastasis
		M_1 Metastasis to one organ system
		M_2 Metastasis to more than one organ system
Cytology cytologic features help assess behavior	Cellular and nuclear morphology are evaluated	Cytology can be superior in evaluation of morphology of hematologic cells, and in some tumors such as poorly granulated mast cell tumors; however, cytologic evaluation lacks ability to assess tumor invasiveness, tumor margins, and presence or absence of vascular invasion.

continued

Table 4-2 Continued

TOOL	MEASURE	EXAMPLE
Histopathology **Histologic grading systems** help correlate morphology with behavior.	Circumscribed or invasive Degree of cellular and nuclear differentiation Mitotic index Presence of lymphatic or blood vascular invasion Metastasis to lymph nodes or other tissues	Mast cell grading systems may help predict behavior of canine mast cell tumors. See example: **Example** Grade I (well differentiated) Grade II (intermediate differentiation) Grade III (anaplastic/poorly differentiated)
Histochemical stains	Differential staining of cells or other tissue components	Toluidine blue procedure stains mast cell granules, helping differentiate poorly differentiated mast cell tumors from other round cell tumors (e.g. lymphoma).
Immunologic (immunohisto-chemical) stains*	Differential staining by antibodies against membrane or cytoplasmic proteins	Cytokeratin Immunostaining Vimentin Immunostaining

Cytokeratin Immunostaining

Carcinoma	Sarcoma	Melanoma
+	−	−

Vimentin Immunostaining

Carcinoma	Sarcoma	Melanoma
-	+	+

*Differential staining may help identify the classification of a poorly differentiated tumor

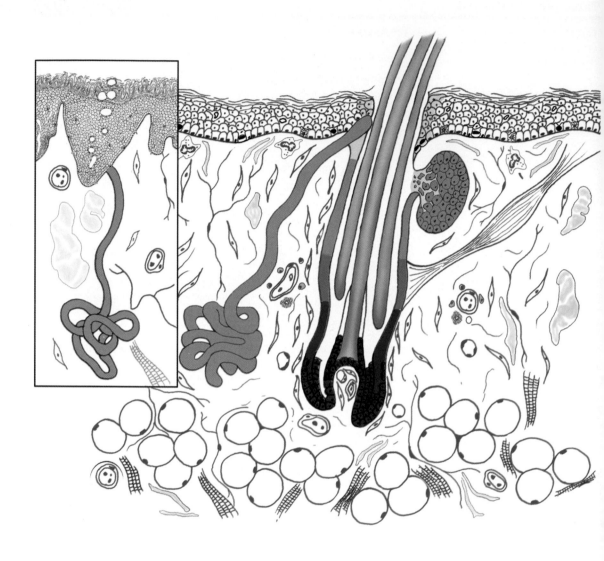

4-8 Schematic representations of the individual cellular components of the skin that can give rise to tumors.

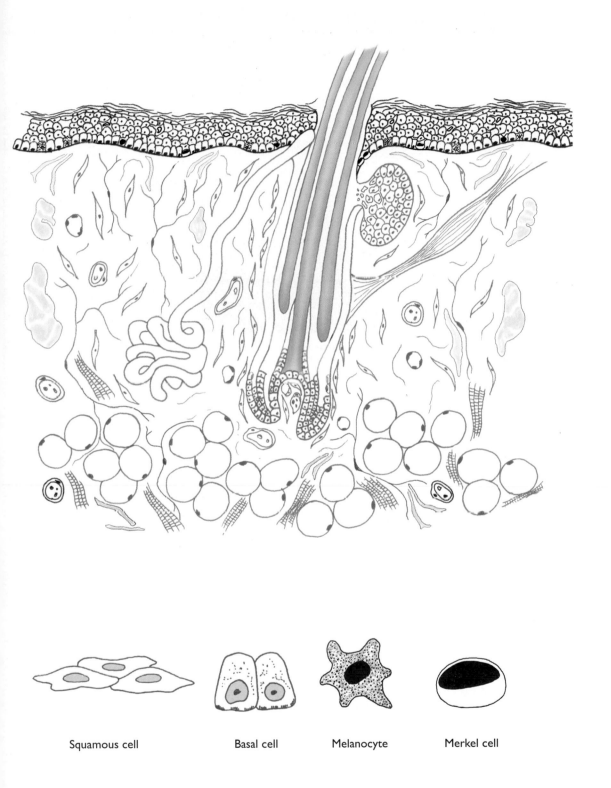

Squamous cell   Basal cell   Melanocyte   Merkel cell

4-9 Schematic representations of the individual cellular components of the epidermis that can give rise to tumors. These same caricatures are placed within the tables in order to help practitioners and residents focus on the cell of origin of the tumors.

Table 4-3 Primary Tumors of the Skin Arising Within the

CELL OF ORIGIN	BENIGN TUMOR	INTERMEDIATE
Squamous cell (Figure 4-10) 	Papilloma (wart) (Figure 4-11) Fibropapilloma 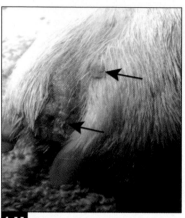 **4-11** Papilloma in interdigital skin of dog. Multiple, raised papules with tan adherent scale-crust (arrows).	Feline fibropapilloma (histologically resembles equine sarcoid) and is associated with papillomavirus infection; may recur. Solar keratosis (Figure 4-12) 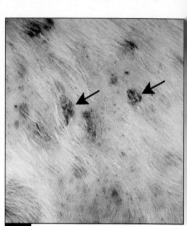 **4-12** Solar dermatitis with solar keratoses on flank in a Bull terrier. Lesions consist of erythema, papules, comedones, crusts, ulcers, and furuncles (due to bacterial infection in dilated plugged follicles). The solar keratoses are depicted by arrows. Treatment with antibiotics prior to biopsy collection helps eliminate bacterial infection thereby distinguishing infection related papules from those associated with epidermal hyperplasia and atypia, a clinical differentiation that can be difficult.
	Pigmented epidermal plaque (epidermodysplasia verruciformis) (pigmented epidermal nevus) (epidermal hamartoma, misnomer)	Associated with papillomavirus infection in dogs. The predisposition to pigmented epidermal plaque may be acquired or inherited immune deficiency to the papillomavirus infection.
	Claw bed (subungual) keratoacanthoma	

Epidermis

MALIGNANT TUMOR	PROGNOSIS	TREATMENT
Squamous cell carcinoma (Figure 4-13) Multicentric carcinoma in situ (Bowen's disease) Basosquamous cell carcinoma  **4-13** **Squamous cell carcinoma** arising on the nasal planum of a cat. The tumor presented as crusting, focal ulceration, and swelling deforming the right side of the planum. Due to local invasion, recurrence of squamous cell carcinomas in this site is common.	Papilloma: Usually good, but some papillomas may not resolve, and a few may undergo malignant transformation. Solar keratoses: Good, but may progress to squamous cell carcinoma. Bowen's disease: Good, but uncommonly invade the dermis. Squamous cell and basosquamous cell carcinoma good if diagnosed early as metastatic rate is usually low and metastases occur late if at all; recurrence is seen in areas where complete excision is difficult (eyelid/nasal planum) or on ventral abdomen where all precancerous actinic damage can not be removed. Higher grade tumors are more likely to metastasize. Claw bed squamous cell carcinomas in cat can be aggressive.	Papilloma: Surgical excision; viral types can spontaneously regress. Solar keratosis: Surgical excision, but other keratoses can develop in adjacent sun-exposed skin. Bowen's disease: Surgical excision if single or small numbers. Topical chemotherapy shows promise.* Squamous cell and basosquamous cell carcinoma: wide surgical excision, radiation therapy, photodynamic therapy (cats) for primary; chemotherapy for metastasis (dogs).* Claw bed tumors usually require digital amputation.
	Good, but some may progress to squamous cell carcinoma.	Surgical excision, but may be too numerous.

Claw bed (subungual) squamous cell carcinoma	Keratoacanthoma:
	    Excellent
	
	Squamous cell carcinoma
	
	    Dog: can metastasize to regional lymph node
	
	    Cat: Frequent lymph node metastasis
		Surgical excision, usually requires digital amputation.

continued

* Treatment modalities may change rapidly in the field of oncology, therefore especially with aggressive, uncommon, or unusual tumor presentations oncology consultation or referral may be advisable.

Table 4-3 Continued

CELL OF ORIGIN	BENIGN TUMOR	INTERMEDIATE
Basal cell (Figure 4-14)	Benign basal cell tumor (basal cell epithelioma) (Figure 4-15) 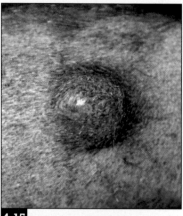 **4-15** **Basal cell tumor** in a cat. Benign tumor that presents as a discrete, frequently pigmented nodule.	
Melanocyte (Figure 4-16) Melanocytes may also be found in the follicular epithelium and dermis)	Melanocytoma (benign melanoma) Rare in cats Melanoacanthoma	
Merkel cell (Figure 4-17)	Merkel cell tumor	

MALIGNANT TUMOR	PROGNOSIS	TREATMENT
Basal cell carcinoma	Benign: Excellent Malignant: Good if diagnosed early as these tumors rarely if ever metastasize. Recurrence is possible with incomplete removal.	Benign: Surgical excision Malignant: Wide surgical excision
Malignant melanoma Rare in cats	Melanocytoma: Good** Malignant melanoma: Poor, often recur and metastasize early to regional lymph nodes and lung in particular. Cats may develop satellite metastases.	Melanocytoma: Wide surgical excision Malignant melanoma: Wide surgical excision of primary. Adjunctive chemotherapy may provide palliation.*
	Good, usually have benign behavior Post surgical recurrence is rare	Surgical excision

* Treatment modalities may change rapidly in the field of oncology, therefore especially with aggressive, uncommon, or unusual tumor presentations oncology consultation or referral may be advisable.
** It can be impossible to histologically differentiate between benign and malignant melanocytic tumors. A small percentage of well-differentiated "histologically benign appearing" melanocytic tumors may metastasize, and are therefore actually malignant tumors.
WHO Histological classification of epithelial and melanocytic tumors of the skin of domestic animals (see references)
WHO Histological classification of mesenchymal tumors of the skin and soft tissues of domestic animals (see references)

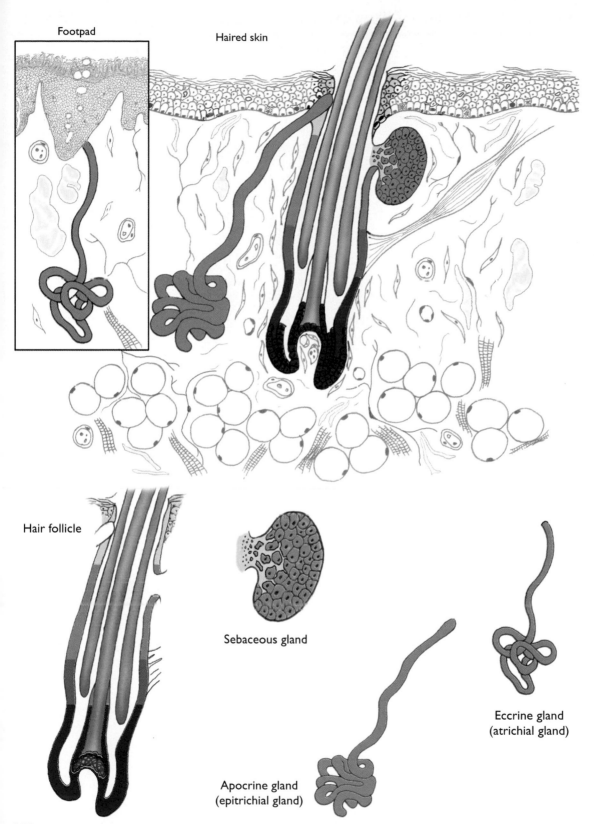

Footpad

Haired skin

Hair follicle

Sebaceous gland

Apocrine gland
(epitrichial gland)

Eccrine gland
(atrichial gland)

4-18 Schematic representations of the individual cellular components of the adnexa that can give rise to tumors. These same caricatures are placed within the tables in order to help practitioners and residents focus on the cell of origin of the tumors.

Table 4-4 Primary Tumors of the Skin Arising in the Adnexa

CELL OF ORIGIN	BENIGN TUMOR	INTERMEDIATE
Hair follicle epithelium (different levels) (Figure 4-19) 	Infundibular keratinizing acanthoma (keratoacanthoma, intracutaneous cornifying epithelioma) (Figure 4-20) 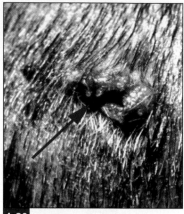 **4-20** Infundibular keratinizing acanthoma in a dog. These tumors are typically concave with a central cavity (arrow) connected to the skin surface. Cutaneous horns may arise from the keratin accumulating in the central cavity. In this case the horn has been lost, exposing the central cavity.	
	Pilomatricoma (pilomatrixoma, necrotizing and calcifying epithelioma of Malherbe)	
	Trichoblastoma (granular, trabecular, ribbon, spindle)	
	Trichoepithelioma	Infiltrative trichoepithelioma Rare
	Tricholemomma Rare	

MALIGNANT TUMOR	PROGNOSIS	TREATMENT
	Excellent; however, in predisposed breeds such as Norwegian elkhound, multiple tumors may develop.	Surgical excision is treatment of choice. If multicentric consider chemotherapy.*
Malignant pilomatricoma (pilomatrix carcinoma)	Benign: Excellent Malignant: Poor, often recur or metastasize to lymph node or lung.	Benign: Surgical excision Malignant: Wide surgical excision. No treatment reported if metastatic.
	Excellent	Surgical excision
Malignant trichoepithelioma Rare	Benign: Excellent; however, in predisposed breeds such as the Basset hound, multiple tumors may develop. Infiltrative: Due to infiltrative nature, may recur. Malignant: The tumor may grow rapidly and may metastasize to regional lymph nodes and lung.	Benign: Surgical excision Infiltrative: Wide surgical excision Malignant: Wide surgical excision
	Excellent	Surgical excision

continued

* Treatment modalities may change rapidly in the field of oncology, therefore especially with aggressive, uncommon, or unusual tumor presentations oncology consultation or referral may be advisable.

Table 4-4 Continued

CELL OF ORIGIN	BENIGN TUMOR	INTERMEDIATE
Sebaceous gland (Figure 4-21)	Sebaceous adenoma (Figure 4-22) 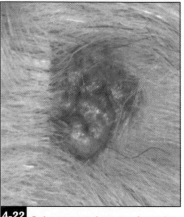 **4-22** **Sebaceous adenoma** from the skin of a poodle. This tumor was one of many nodular, hairless masses.	Sebaceous epithelioma
	Meibomian adenoma (gland or ductal)	Meibomian epithelioma
Modified sebaceous gland	Hepatoid gland adenoma (perianal, circumanal gland adenoma) (Figure 4-23) 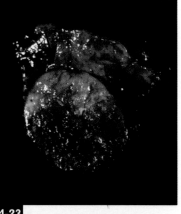 **4-23** **Section of hepatoid (perianal) gland adenoma.** Benign mass with circumscribed mode of growth. This mass is completely excised. Note circumscribed margin (arrows).	Hepatoid gland epithelioma (perianal, circumanal gland epithelioma)

MALIGNANT TUMOR	PROGNOSIS	TREATMENT
Sebaceous carcinoma Rare	Adenoma: Excellent Epithelioma: Usually good, but may recur with incomplete excision; rare metastasis to regional lymph nodes. Carcinoma: Usually these tumors are only locally invasive. Infrequently metastasize to lymph nodes, and rarely to lungs.	Surgical excision for all types, with wide excision for carcinoma.
Meibomian carcinoma	Same as for sebaceous gland tumors	Same as for sebaceous gland tumors
Hepatoid gland carcinoma (perianal, circumanal gland carcinoma)	Adenoma: Excellent Epithelioma: Good, but may recur with incomplete excision. Carcinoma: Fair to poor. May metastasize to regional lymph nodes, lungs, and other organs.	All types: Surgical excision is treatment of choice. Adenomas are often androgen responsive so castration may be useful; large tumors or poor surgical candidates may benefit from radiation therapy.*

continued

* Treatment modalities may change rapidly in the field of oncology, therefore especially with aggressive, uncommon, or unusual tumor presentations oncology consultation or referral may be advisable.

Table 4-4 Continued

CELL OF ORIGIN	BENIGN TUMOR	INTERMEDIATE
Apocrine gland (Figure 4-24) (epitrichial gland)	Apocrine adenoma, simple, complex, mixed	
	Apocrine ductal adenoma	
	Ceruminous adenoma, simple, complex, mixed	
	Anal sac gland adenoma (adenoma of the apocrine glands of anal sac)	
Eccrine gland (atrichial gland) (Figure 4-26)	Eccrine adenoma	

MALIGNANT TUMOR	PROGNOSIS	TREATMENT
Apocrine carcinoma, simple, complex, mixed (Figure 4-25) Rare  **4-25** **Apocrine adenocarcinoma** in a cocker spaniel. Large, nodular mass with ulceration. This mass was "fixed" to surrounding tissue due to invasion of cells into the bordering subcutis and musculature.	Adenoma: Excellent Carcinoma: Tumors that have an "inflammatory" clinical appearance or those that are poorly differentiated usually have a poor prognosis as many metastasize to lymph nodes and lungs. Nodular types, or those that have complex or mixed morphology tend to be slower growing and less likely to metastasize.	Surgical excision. Behavior of apocrine gland carcinomas is similar to mammary gland carcinomas, so therapy for both is similar.
Apocrine ductal carcinoma	Adenoma: Excellent Carcinoma: Fair to good as metastases are uncommon.	Adenoma: Surgical excision Carcinoma: Wide surgical excision
Ceruminous gland carcinoma, simple, complex, mixed	Adenoma: Good, but non-pedunculated tumors may be difficult to completely excise and may recur. Carcinoma: Poor due to local invasion. There may be metastasis to regional lymph nodes.	Adenoma: Surgical excision, may require ablation of ear. Carcinoma: Surgical ablation of ear canal for primary. Tumor may extend into middle ear. In cats the tumors may occur bilaterally.*
Anal sac gland adenocarcinoma (adenocarcinoma of the apocrine glands of the anal sac)	Adenoma: Good Carcinoma: Poor, locally invasive, metastasize early to regional lymph nodes. Some metastasize widely.	Adenoma: Surgical excision of tumor and anal sac Carcinomas: Wide surgical excision
Eccrine adenocarcinoma	Adenoma: Good, but occurs on footpad. Carcinoma: Fair to good as metastasis are infrequent.	Adenoma: Surgical excision Carcinoma: Wide surgical excision

* Treatment modalities may change rapidly in the field of oncology, therefore especially with aggressive, uncommon, or unusual tumor presentations oncology consultation or referral may be advisable.
WHO Histological classification of epithelial and melanocytic tumors of the skin of domestic animals (see references)
WHO Histological classification of mesenchymal tumors of the skin and soft tissues of domestic animals (see references)

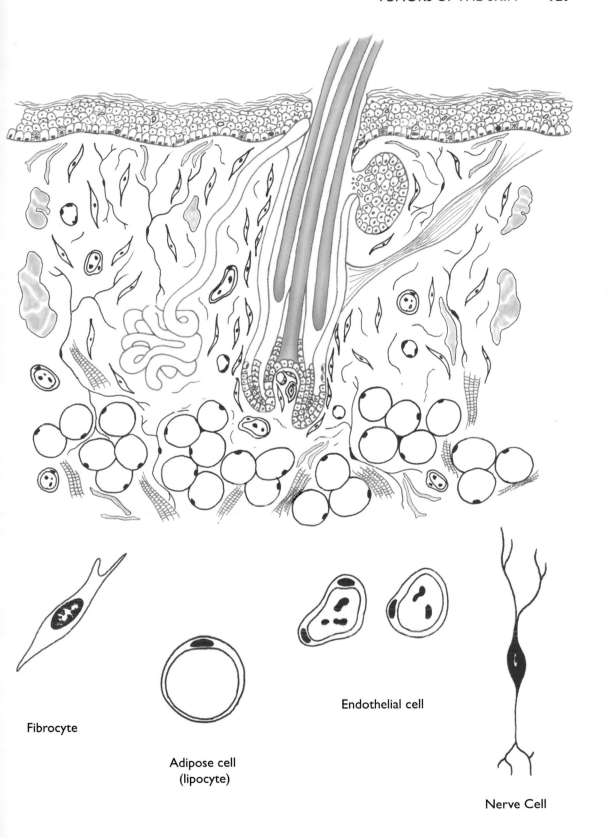

Fibrocyte

Adipose cell
(lipocyte)

Endothelial cell

Nerve Cell

4-27 Schematic representations of the individual cellular components of the dermis and subcutis that can give rise to tumors. These same caricatures are placed within the tables in order to help practitioners and residents focus on the cell of origin of the tumors.

Table 4-5 Primary Tumors of the Skin Arising in the Dermis,

CELL OF ORIGIN	BENIGN TUMOR	INTERMEDIATE
Fibrocyte (Figure 4-28) (location in dermis) 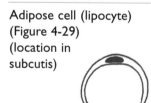	Fibroma	
	Myxoma	
Primitive mesenchymal cell (location in dermis)		
Adipose cell (lipocyte) (Figure 4-29) (location in subcutis)	Lipoma (Figure 4-30) 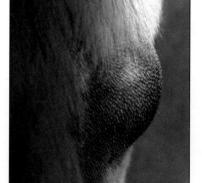	Infiltrative lipoma (infiltrative, but does not metastasize)

4-30 **Lipoma** located on the limb of a dog. This solitary mass is in the subcutis and is not attached to the skin or deeper tissues.

Vasculature, Nerve Tissue, or Subcutis

MALIGNANT TUMOR	PROGNOSIS	TREATMENT*
Fibrosarcoma Feline multicentric fibrosarcoma (Feline post vaccinal fibrosarcoma if there is evidence of vaccine association)	Fibroma: Excellent Fibrosarcoma: Small, slowly growing tumors in sites amenable to wide surgical excision have a better prognosis. In dogs, recurrence is not uncommon; metastases may occur in about 10-15% of cases. However, recent information suggests that poorly differentiated fibrosarcomas may have a metastatic rate of up to 50%. In cats, recurrence is extremely common, metastases occur in about 10-15% of cases.	Fibroma: Surgical excision Fibrosarcoma: Wide surgical excision, possibly in combination with radiation therapy* Feline post vaccinal fibrosarcomas often recur. Metastasis appears to increase with prolonged survival.
Myxosarcoma	Guarded as both tumor types tend to recur but metastasize rarely.	Wide surgical excision. Amputation may be curative in those cases where there has been repeated recurrence.*
Malignant fibrous histiocytoma (giant cell tumor of soft tissue and extraskeletal giant cell tumor)	"Malignant fibrous histiocytoma" probably consists of a variety of different tumor types that are difficult to differentiate histologically. Prognosis may depend on identifying the cell of origin. In general these tend to recur.	Wide surgical excision possibly in combination with radiation therapy.*
Liposarcoma Rare	Lipoma: Excellent Infiltrative lipoma: Guarded as recurrence is frequent. Liposarcoma: Good, most tumors do not recur, and metastasis is rare.	Lipoma: Surgical excision if large size becomes a problem Infiltrative lipoma: Wide, aggressive, early excision** Liposarcoma: Wide surgical excision

* Note: because the behavior of many different types of soft tissue sarcomas is locally invasive with low metastatic potential, there is a similar treatment protocol for these various tumors by some oncologists. The treatment protocol currently used at the institution of one of the authors (Bettenay) includes the widest surgical excision possible, radiation therapy, ± chemotherapy.
** Note: treatment modalities may change rapidly in the field of oncology, therefore especially with aggressive, uncommon, or unusual tumor presentations oncology consultation or referral may be advisable.   continued

Table 4-5 Continued

CELL OF ORIGIN	BENIGN TUMOR	INTERMEDIATE
	Angiolipoma	
Endothelial (Figure 4-31) (location in vessels)	Hemangioma	Kaposi-like vascular tumor Rare and controversial entity
	Lymphangioma	

MALIGNANT TUMOR	PROGNOSIS	TREATMENT
	Excellent	Surgical excision
Hemangiosarcoma (Figures 4-32 and 4-33)	Hemangioma: Usually good if can be completely excised.	Hemangioma: Surgical excision Kaposi-like vascular tumor: surgical excision
	Kaposi-like vascular tumor: good if single, but if multiple recurrences are more common.	Hemangiosarcoma: Wide surgical excision. Adjunctive Chemotherapy*
	Hemangiosarcoma: It is important to differentiate primary from metastatic hemangiosarcoma through staging for possible primary tumors in the spleen, liver or heart. Primary cutaneous hemangiosarcomas are less aggressive than visceral tumors (fewer metastases).	
	Large or multifocal vascular tumors can be associated with hemostatic abnormalities.	

4-32 Hemangiomatous tumors on the caudal thigh of a whippet. These tumors are solar associated and occur in non pigmented sparsely haired skin.

4-33 Ventral abdomen of another whippet. There are multiple vascular tumors in the skin. Small superficial tumors are likely primary. However, the deeper nodule with bluish appearance (arrow) is likely metastatic.

Lymphangiosarcoma	Lymphangioma: Poor, as tumor frequently recurs.	Both types: Surgical excision if possible. Radiation therapy may be useful.*
	Lymphangiosarcoma: Poor as tumor frequently recurs. Metastasis is rare.	
Feline ventral abdominal angiosarcoma	Guarded to poor as these tumors frequently recur, but rarely metastasize.	Surgical excision if diagnosed early and if possible.

continued

* Note: treatment modalities may change rapidly in the field of oncology, therefore especially with aggressive, uncommon, or unusual tumor presentations oncology consultation or referral may be advisable.

Table 4-5 Continued

CELL OF ORIGIN	BENIGN TUMOR	INTERMEDIATE
Presumptive Schwann cell origin (Figure 4-34) (location in nerve)	Granular cell tumor	
Perineural fibroblast or Schwann cell or both	Benign peripheral nerve sheath tumor (neurofibroma, schwannoma)	
Pericyte, peripheral nerve sheath, or perineural fibroblast (exact cell of origin undetermined)		Hemangiopericytoma

MALIGNANT TUMOR	PROGNOSIS	TREATMENT
	Good	Surgical excision
Malignant peripheral nerve sheath tumor (neurofibrosarcoma, malignant schwannoma).	Benign: Good, but some may recur after surgical excision. Malignant: Guarded as tumors tend to recur.	Benign: Surgical excision Malignant: Wide surgical excision*
	Guarded as recurrence is frequent (about 30% of cases), and with each recurrence the tumor becomes more aggressive.	Wide surgical excision. Amputation can be curative in cases with repeated recurrences.*

* Note: because the behavior of many different types of soft tissue sarcomas is locally invasive with low metastatic potential, there is a similar treatment protocol for these various tumors by some oncologists. The treatment protocol used at the institution of one of the authors (Bettenay) includes the widest surgical excision possible, radiation therapy, ± chemotherapy.
WHO Histological classification of epithelial and melanocytic tumors of the skin of domestic animals (see references)
WHO Histological classification of mesenchymal tumors of the skin and soft tissues of domestic animals (see references)

Table 4-6
Examples of Grading Systems for Canine Mast Cell Tumors

	Grading Systems*		
	Bostock	**Hottendorf**	**Patnaik**
Grade of Tumor			
Low grade	Grade 3	Well-differentiated	Grade I
Intermediate grade	Grade 2	Intermediate differentiation	Grade II
High grade	Grade 1	Anaplastic	Grade III

*Multiple grading systems for canine mast cell tumors have been developed. Notice the numerical system of Bostock and Patnaik are opposite each other. Make certain you know which system is used by the pathologist reading your samples!

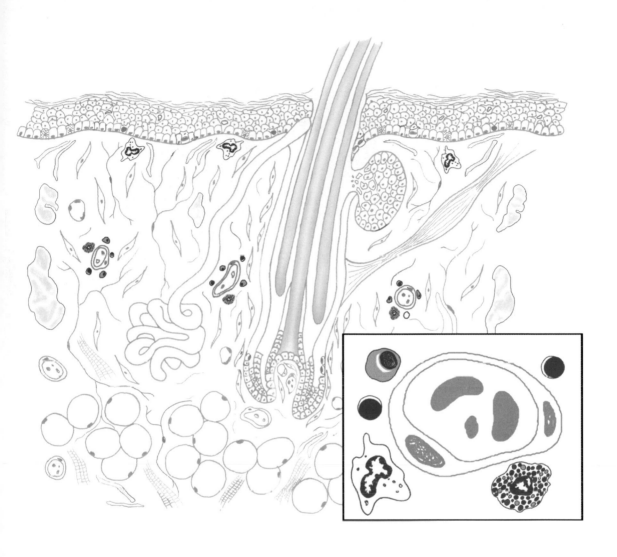

Histiocyte
(Dendritic antigen
presenting cells)

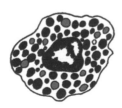

Mast cell

Lymphocyte

Plasma cell

4-35 Schematic representations of the individual cellular components of the bone marrow derived cells that may give rise to tumors. These same caricatures are placed within the tables in order to help practitioners and residents focus on the cell of origin of the tumors.

Table 4-7 Tumors of the Skin Arising from Bone Marrow Derived Cells

CELL OF ORIGIN	BENIGN TUMOR	INTERMEDIATE
Histiocyte (Figure 4-36) (Dendritic antigen presenting cells) (origin in epidermis and dermis)	Canine cutaneous histiocytoma (Figure 4-37)	Canine reactive histiocytosis considered reactive, not a true neoplasm (cutaneous and systemic histiocytosis) (Figure 4-38)

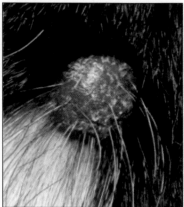

4-37 **Canine cutaneous histio-cytoma** from a young dog. Histiocytomas are typically solitary hairless masses and are regarded as a reactive proliferation of Langerhans cells, rather than a neoplasm. Spontaneous regression usually occurs.

4-38 **Canine reactive histiocytosis** from a mature dog. The dog presented with plaques and nodules, arising on the dorsal muzzle and lateral nasal planum, trunk and limbs. Canine reactive histiocytosis often involves the nasal mucosa as it did in this dog.

(also known as "discrete cell tumors" or "round cell tumors")

MALIGNANT TUMOR	PROGNOSIS	TREATMENT
Malignant histiocytosis	Cutaneous histiocytoma: Good as most cases either spontaneously regress or are cured by excision.	Cutaneous histiocytoma: Surgical excision usually curative; however histiocytomas usually spontaneously regress.
	Reactive histiocytosis: Guarded. Often waxing and waning disease that is progressive and becomes refractory to therapy although occasional cases may spontaneously regress.	Reactive histiocytosis: Single mass may be amenable to surgical excision. Treatment of multicentric masses in systemically healthy dog is controversial.*
	Malignant: Poor, usually involves many organs.	Malignant: No known successful treatment. Supportive treatment may be as important in chemotherapy in prolonging survival.*

continued

* Treatment modalities may change rapidly in the field of oncology, therefore especially with aggressive, uncommon, or unusual tumor presentations oncology consultation or referral may be advisable.

Table 4-7 Continued

CELL OF ORIGIN	BENIGN TUMOR	INTERMEDIATE
Mast cell* (Figure 4-39) (origin in dermis)	Feline well differentiated mast cell tumor (Figure 4-40) 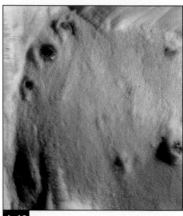 **4-40** **Mast cell tumors** in a cat. Multiple, non-pruritic papules and ulcerated nodules are in the skin of the trunk. (Courtesy of Dr. Ken Mason).	
	Feline histiocytic mast cell tumor	
	Canine well differentiated mast cell tumor (Grade 1, Grade 3)	Canine intermediate grade mast cell tumor (Grade II, Grade 2)

* Mast cell tumors do not fit within strict benign and malignant categories. In general feline mast cells tumors tend to have a benign biologic behavior. The histiocytic variety has been reported in young cats and has exhibited spontaneous regression. Poorly differentiated feline mast cell tumors may be more likely to recur. All grades of mast cell tumors in dogs should be considered potentially aggressive. Some authors believe that "well-differentiated" canine mast cell tumors are less aggressive than tumors in "intermediate" and "poorly" differentiated grades, but other authors disagree.

MALIGNANT TUMOR	PROGNOSIS	TREATMENT
Feline poorly differentiated mast cell tumor	Good as most feline mast cell tumors have a benign behavior.	Wide surgical excision for a single lesion. In cats with multiple lesions, glucocorticoids may be used at anti-inflammatory doses. Cats should be evaluated for systemic mast cell tumors by complete clinical staging.
	Good	Can spontaneously regress
Canine poorly differentiated mast cell tumor (Grade III, Grade 1)	Prognosis varies with grade and stage, but all mast cell tumors should be considered potentially malignant. However, usually poorly differentiated tumors recur more frequently than well differentiated tumors. Grade I usually do not recur, and 3 year survival rate is 90%. Grade II recurrence is common, and 3 year survival rate is 55%. Grade III recurrence is common, and 3 year survival rate is 10-15%.	Wide and deep surgical excision if there is no evidence of lymph node involvement or systemic spread.*

continued

* Treatment modalities may change rapidly in the field of oncology, therefore especially with aggressive, uncommon, or unusual tumor presentations oncology consultation or referral may be advisable.

Table 4-7 Continued

CELL OF ORIGIN	BENIGN TUMOR	INTERMEDIATE
Lymphocyte (Figure 4-41) (epidermis or dermis)		
Plasma cell (Figure 4-43) (dermis)	Plasmacytoma (extramedullary plasmacytoma) Uncommon in dogs and rare in cats	Tumors with amyloid or in subcutis may be more likely to recur.

MALIGNANT TUMOR	PROGNOSIS	TREATMENT
Malignant lymphoma Uncommon Epidermal (epitheliotropic lymphoma, Woringer-Kolopp disease, pagetoid reticulosis, mycosis fungoides) (Figure 4-42) Dermal (non epitheliotropic lymphoma, dermal lymphosarcoma) Angiocentric (lymphomatoid granulomatosus) rare	Epidermal: Poor. Clinical course may be prolonged if diagnosed early, but disease is progressive. Ulceration leads to secondary infection and sepsis. Spread to lymph nodes and viscera is typically late. Dermal: Poor Angiocentric: Poor, usually also involves lung	Epidermal: Therapy is palliative and does not affect the survival time. Death is often due to euthanasia. Dermal: Chemotherapy, often unrewarding* Angiocentric: too few cases reported to judge*

4-42 **Epitheliotropic lymphoma** (mycosis fungoides) in a dog. Severe erythema, scale-crust and ulceration affect the feet. These lesions developed after several years of topical glucocorticoid responsive contact allergic dermatitis. (Courtesy of Dr. Helen Power.)

Poorly differentiated plasmacytoma	Good. In dogs, usually have benign behavior, but those with amyloid may recur. Local recurrence and metastasis are rare. Those in oral cavity or subcutis may be more aggressive. Too few studies reported in cats to judge prognosis. Rarely reported with multiple myeloma.	Surgical excision

* Treatment modalities may change rapidly in the field of oncology, therefore especially with aggressive, uncommon, or unusual tumor presentations oncology consultation or referral may be advisable.
WHO Histological classification of epithelial and melanocytic tumors of the skin of domestic animals (see references)
WHO Histological classification of mesenchymal tumors of the skin and soft tissues of domestic animals (see references)

Table 4-8 Cysts (any sac-like structure containing liquid or

COMPONENT GIVING RISE TO CYST	NAME OR TYPE OF CYST	HISTOLOGIC FEATURES
Epidermal	Epidermal inclusion cyst* (epidermoid cyst, epidermal cyst) (Figure 4-44) 4-44 Six month old Corgi cross. Multiple, alopecic, cystic nodules arose in a linear pattern along the inner thigh and ventral abdomen.	Usually unilocular cyst lined by stratified squamous epithelium of the epidermis (basal, spinous, granular, and cornified layers) and containing stratum corneum.
Adnexal Follicular	Infundibular*	Usually unilocular cyst lined by stratified squamous epithelium of the follicular infundibular epithelium (basal, spinous, granular, and cornified layers) and containing stratum corneum.
	Isthmus	Cyst lined by stratified squamous epithelium without granular layer (isthmus or inferior segments of follicle).
	Panfollicular (trichoepitheliomatous)	Stratified squamous epithelium consistent with infundibular and isthmus cysts. May also contain small basal cells with abrupt cornification and intracytoplasmic trichohyalin granules.
	Dilated pore (pore of Winer)	Unilocular cyst lined by hyperplastic stratified squamous epithelium that has rete ridges and a large pore communicating with the epidermis. The lumen contains compact cornified cells that may extend through the opening. The granular layer is prominent. A cyst found in cats.

*Differentiation of these two cysts may not be possible histologically. They arise in stratified squamous keratinizing epithelium either of epidermis or follicular infundibular epithelium.

semisolid material)

LOCATION IN SKIN	PROGNOSIS	TREATMENT
Dermis or subungual If cyst is large, may extend into subcutis.	Excellent	Surgical excision
Dermis or if cyst is large, may extend into the subcutis.	Excellent	Surgical excision
Dermis	Excellent	Surgical excision
Dermis	Excellent	Surgical excision
Dermis A large pore through the epidermis connects the cyst with the surface of the skin.	Excellent	Surgical excision

continued

Table 4-8 Continued

COMPONENT GIVING RISE TO CYST	NAME OR TYPE OF CYST	HISTOLOGIC FEATURES
Epidermal and adnexal	Dermoid (dermoid sinus)	Congenital cyst lined by stratified squamous epithelium of epidermis and containing dermal and adnexal structures in the cyst wall, and hair and cornified cells in the lumen. Can have pore communicating with epidermis.
Sebaceous duct	Sebaceous duct cyst	Cyst lined by sebaceous ductal squamous epithelium and surrounded by hyperplastic sebaceous glands.
Apocrine	Single	Cyst lined by apocrine epithelium and containing clear secretion.
	Multiple (apocrine cystomatosis)	Multiple cysts lined by apocrine epithelium and containing clear secretion. The cysts are in multiple sites in the skin.
Other	Ciliated	Cyst lined by cuboidal to columnar epithelium with or without goblet cells. Probable developmental anomaly of thyroglossal ducts or respiratory tract. Usually found in neck of cats.

LOCATION IN SKIN	PROGNOSIS	TREATMENT
Dermis A pore may connect cyst through epidermis to surface of skin.	Varies with depth of cyst. Some cysts may attach to the dura mater of the spinal canal.	Varies with depth of cyst. Some cysts can attach to the dura mater of the spinal canal.
Dermis	Excellent	Surgical excision
Dermis	Excellent	Surgical excision
Dermis	Varies with extent of involvement	Varies with extent of involvement
Dermis	Excellent	Surgical excision

WHO Histological classification of epithelial and melanocytic tumors of the skin of domestic animals (see references)
WHO Histological classification of mesenchymal tumors of the skin and soft tissues of domestic animals (see references)

Table 4-9 Hamartomas*(a disturbance of tissue growth where the cells

COMPONENT GIVING RISE TO HAMARTOMA	NAME OR TYPE OF HAMARTOMA	HISTOLOGIC FEATURES
Epidermal	Epidermal hamartoma** (Figure 4-45) 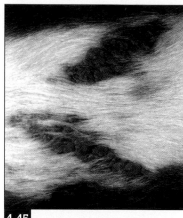 **4-45** **Epidermal hamartomas** in an 8 month old Basset hound. Multiple hairless plaques in linear arrays arose in the skin. The lesions were present at birth.	Redundant epidermis (folded, papillary epidermis, normal in morphology, but excessive in quantity, sometimes following linear configurations in the skin, and usually hairless). The degree of pigment varies with that in the epidermis in which the hamartoma arises.
Follicular	Follicular hamartoma (Figure 4-46) 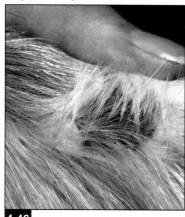 **4-46** **Follicular hamartoma** in a dog. The usual clinical presentation is a haired nodule with widely spaced enlarged follicular openings, containing large primary hairs. (Courtesy of Dr. Ralf Mueller.)	Nodular or plaque-like areas of enlarged follicles with a variable quantity of collagen and normal or mildly hyperplastic sebaceous glands. The deep portions of follicles are embedded deeply in the subcutis. These lesions are best visualized with incisional biopsy techniques that include normal skin (normally sized follicles) with which to compare the enlarged follicles.
Sebaceous	Sebaceous hamartoma	Increased numbers of mature sebaceous glands that may or may not be associated with a follicle. Usually bordered by collagen.

* The lesions in this table are of presumed congenital origin.
** There has been confusion regarding type of lesions considered to represent epidermal hamartomas. One type of lesion is considered by the authors to be a hamartoma or congenital malformation. However, the authors believe the other lesion (pigmented epidermal plaque, see tumors of the epidermis) has been incorrectly termed "epidermal hamartoma". The pigmented epidermal plaque is due to an apparent immune deficiency (acquired or possibly inherited) predisposing to infection with a papillomavirus. Selected breeds of dogs (pug dogs and miniature schnauzers) are predisposed to infection and lesion development.

of a circumscribed area surpass the surrounding area)

LOCATION IN SKIN	PROGNOSIS	TREATMENT
Epidermis	Depends on extent of involvement	Depends on extent of involvement
Dermis and panniculus	Prognosis usually good, but rarely follicular or collagenous hamartoma have recurred and progressed, despite benign histologic appearance.	Surgical excision
Dermis	Excellent	Surgical excision

continued

Table 4-9 Continued

COMPONENT GIVING RISE TO HAMARTOMA	NAME OR TYPE OF HAMARTOMA	HISTOLOGIC FEATURES
Apocrine	Apocrine	Focal increase in number of apocrine glands.
Fibroadenxal	Fibroadenxal (synonyms: adnexal nevus, adnexal dysplasia, folliculosebaceous hamartoma)*	Usually nodular area of disorganized large adnexal units (follicles and sebaceous glands embedded within abundant collagen). There is no connection with the overlying epidermis, and there may be foci of inflammation.
Collagenous	Collagenous hamartoma (Figure 4-47)	Nodular poorly defined accumulation of excessive collagen that may result in slight elevation of epidermis and loss or displacement of adnexa. One report of extensive infiltration into surrounding tissue in three dogs.

4-47 **Collagenous hamartoma** in a dog. The alopecia is due to displacement and loss of the adnexa, common with this hamartoma.

* Some of these lesions are likely congenital, but others may be secondary to trauma.

LOCATION IN SKIN	PROGNOSIS	TREATMENT
Dermis and/or subcutis	Excellent	Surgical excision
Dermis and subcutis	Excellent	Surgical excision
Superficial dermis	Prognosis is usually good, but rarely collagenous or follicular hamartoma have recurred, despite benign histologic appearance.	Surgical excision

continued

WHO Histological classification of epithelial and melanocytic tumors of the skin of domestic animals (see references)
WHO Histological classification of mesenchymal tumors of the skin and soft tissues of domestic animals (see references)

Table 4-9 Continued

COMPONENT GIVING RISE TO HAMARTOMA	NAME OR TYPE OF HAMARTOMA	HISTOLOGIC FEATURES
Nodular dermatofibrosis	Nodular dermatofibrosis of the German shepherd* (Figure 4-48) 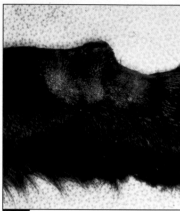 **4-48** **Nodular dermatofibrosis** in a female German Shepherd dog. The hair over the mass has been shaved to facilitate visualization. The collagen proliferation occurs deep in the dermis and subcutis, giving rise to firm, deeply attached nodules that are typically haired. (Courtesy of Dr. Barbara Atlee.)	Multifocal, nodular proliferation of collagen in dermis and subcutis. Adnexa are normal or hyperplastic. Lesions of nodular dermatofibrosis have been associated with renal adenomas and carcinomas, and uterine and vaginal leiomyomas. Most often seen in female German shepherds, but can occur in other breeds and sexes.
Vascular	Scrotal vascular hamartoma	Plaque-like area of progressive, poorly circumscribed, vascular proliferation that can be misdiagnosed as hemangiosarcoma. Proliferative vessels range from arteries to capillaries. A lesion of dogs.

* May more accurately be considered an inherited condition in the German shepherd dog.

LOCATION IN SKIN	PROGNOSIS	TREATMENT
Dermis and subcutis	Nodular dermatofibrosis is a benign lesion in the skin, but is typically multifocal and serves as a marker for renal carcinomas and uterine leiomyomas.	Depends on extent of involvement
Dermis of pigmented scrotal skin	Prognosis usually good, but often is poorly circumscribed.	Surgical excision

WHO Histological classification of epithelial and melanocytic tumors of the skin of domestic animals (see references)
WHO Histological classification of mesenchymal tumors of the skin and soft tissues of domestic animals (see references)

Table 4-10 Other Tumor-like Lesions

COMPONENT GIVING RISE TO TUMOR-LIKE LESION	NAME OR TYPE OF LESION	HISTOLOGIC FEATURES
Epidermal	Cutaneous horn (Figure 4-49) **4-49 Cutaneous horn** in a Bull terrier with solar dermatitis. The cutaneous horn has arisen in a solar keratosis. This animal also has multiple squamous cell carcinomas (not seen in this photograph).	Horn-like projection of stratum corneum that may arise from hyperplastic epidermis, or from benign or less likely malignant epidermal or follicular neoplasms.
Adnexal	Sebaceous gland hyperplasia	Focal or multifocal nodular accumulations of hyperplastic sebaceous glands, often occurring in old dogs.
Epidermal and dermal	Fibroepithelial polyp (skin tag, acrochordon)	Focal or multifocal polypous mass of collagen covered by epidermis. Adnexa are usually sparse or not present.
	Fibropruritic nodule	Focal or multifocal papillary epidermal hyperplasia covered with compact stratum corneum and resting on a core of proliferative fibroblasts and collagen (fibroplasia) that may displace or replace adnexa.
Dermal/Subcutaneous	Nodular fasciitis	Nodular lesion that is poorly understood. Usually considered to be inflammatory, but there are histologic features suggestive of fibrosarcoma (locally infiltrative bundles and whorls of fibroblasts and fibrocytes—some with mitotic figures—intermixed with lymphocytes, plasma cells, and macrophages). A lesion of dogs.

LOCATION IN SKIN	PROGNOSIS	TREATMENT
Epidermis	Prognosis depends on lesion that underlies the cutaneous horn.	Surgical excision
Dermis	Excellent	Surgical excision
Polypoid mass that protrudes above epidermal surface, but contains epidermal and dermal tissues.	Excellent	Surgical excision
Epidermis and dermis	Lesion is benign and prognosis is good if single, but lesions may be multifocal.	Surgical excision of single or small number of lesions
Subcutis	Prognosis is usually good if complete excision is possible.	Surgical excision

continued

Table 4-10 Continued

COMPONENT GIVING RISE TO TUMOR-LIKE LESION	NAME OR TYPE OF LESION	HISTOLOGIC FEATURES
Melanocytic	Melanocytic hyperplasia (lentigo, lentigo simplex)	Plaques of epidermal hyperplasia with increased numbers of melanocytes in the basal layer. The adjacent keratinocytes often contain increased melanin pigment granules, and the dermis has increased numbers of pigment containing macrophages.
Other Lipid containing macrophages	Xanthogranuloma (xanthoma)	Focal or multifocal nodular or plaque-like area of multivac-uolated (foamy) macrophages. May be idiopathic or associated with abnormalities of cholesterol or triglycerides.
Nerve bundles and fibroblasts	Traumatic neuroma (amputation neuroma, tail dock neuroma)	Painful, often hyperpigmented alopecic lesion associated with excessive and disorganized bundles of myelonated axons embedded within collagenous stroma, and caused by surgical or traumatic resection of a nerve. A lesion usually seen in dog, especially associated with tail docking procedures.

LOCATION IN SKIN	PROGNOSIS	TREATMENT
Epidermis	Excellent	Do not require therapy
Dermis	Excellent if idiopathic or if underlying cause can be corrected.	Surgical excision and correcting underlying cause
Dermis	Excellent if surgical excision possible.	Surgical excision

WHO Histological classification of epithelial and melanocytic tumors of the skin of domestic animals (see references)
WHO Histological classification of mesenchymal tumors of the skin and soft tissues of domestic animals (see references)

Section 5

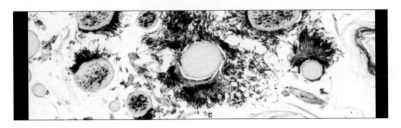

Laboratory Techniques
for Tissue Processing and Staining

5-1 Photograph of inking the margin of a sample. The ink is applied evenly to the surgically excised surface of the tissue and allowed to air dry (until the ink no longer appears shiny) prior to returning the sample to the formalin. The tissue is handled very gently and not squeezed.

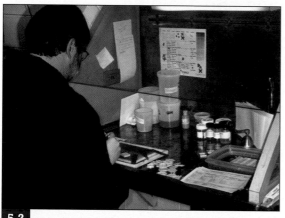

5-2 Photograph illustrates a fume ventilating hood. Formalin is a hazardous substance, and the hood ventilates the hazardous fumes protecting laboratory personnel.

5-3 Photograph illustrates numerous small openings in tissue cassette. The openings allow the processing fluids to contact the tissue.

Introduction

Samples must be processed, sectioned, and stained before histologic sections can be evaluated by a pathologist and a histopathologic diagnosis made. Most of these procedures are performed in the first 24 to 48 hours after the sample is received at the laboratory. Some of the procedures are time consuming and require special equipment and reagents. This section describes this laboratory process for individuals interested in learning about the basics of sample processing, sectioning, and staining.

Sample Arrival at the Laboratory

Packages are opened and samples and paperwork are evaluated, dated, and assigned a reference number. The paperwork and samples are numbered identically to ensure the sample bottle can be matched to paperwork should they become prematurely separated or should collection of additional tissues from the bottle become necessary at a later date. If other samples (for cytology or microbiology etc.) are included in the submission, they are also dated and labeled with the reference number).

Gross Evaluation and Trimming of Sample
Sample and Request Review

✓ The laboratory technician verifies that the paperwork and samples are labeled with the same reference number. The technician reviews the paperwork for special requests (*e.g.*, crusts submitted in tissue paper). Practitioners may have identified, or marked, samples with colored dyes (Appendix B) or suture placed through the nonlesional portion (*e.g.*, the normal end of an elliptical incisional sample), to specify differences in anatomic location on the animal, disease processes, or to distinguish normal skin from lesional skin. The technician uses this information when preparing the samples for further processing. Laboratory technicians may also mark samples with colored dyes to identify samples for the pathologist (*e.g.*, pairs of a bisected sample), and to identify the margins of a nodule (Figure 5-1) so the pathologist can evaluate for completeness of removal.

Sample Trimming

✔ Formalin, the standard fixative, is a hazardous chemical, thus the technician handles the samples in a fume-ventilating hood (Figures 5-2). The samples are trimmed and placed into small plastic or metal tissue cassettes that have numerous small openings that allow free flow of processing chemicals, but prevent the sample from being released (Figure 5-3). The tissue cassettes (Appendix B) are labeled with the sample reference number (Figure 5-4), and additional numbers or letters to identify the specially marked samples (*e.g.*, to differentiate two samples from one animal, such as one nodule from the head and a second nodule from the toe).

☞ At the discretion of the laboratory technician, punch and incisional (elliptically shaped) biopsy samples may be embedded in total or bisected. Elliptical samples usually are trimmed longitudinally, (along the long axis of the sample) (Figure 5-5). Laboratory protocols are used for biopsy sample sectioning because the clinical morphologic features of samples are destroyed by fixation. The alteration of the color, texture, and shape of samples makes visual orientation difficult (Figure 5-6). The technique of drawing a thin line on the surface of the sample with a fine tipped permanent marking pen (see Figure 1-1) enables the technician to orient the sample correctly (in the manner the practitioner desires).

When trimming larger samples, such as excisional samples of suspected tumors, the laboratory protocols established by the processing laboratory ensure that representative areas of the sample are collected. These protocols ensure that deep and lateral margins are collected so the pathologist can evaluate adequacy of excision. Residual tissue is returned to the numbered bottle and stored for a variable time frame (depending on the individual laboratory protocol, typically about one month) should further sections be required.

Tissue Demineralization

Bone or heavily mineralized areas in samples (*e.g.*, in amputated digits, calcinosis cutis with bony metaplasia, mineralized areas in granulomas) require demineralization, also termed, "decalcification". This

5-4 Photograph of multiple cassettes ready for processing.

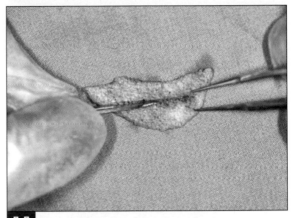

5-5 Photograph of trimming an elliptically shaped sample longitudinally (along the line drawn by the practitioner).

5-6 Photograph of a punch biopsy sample of skin after fixation in formalin. Note how formalin fixation reduces clarity of the clinical lesion.

process removes the minerals (thus the hardness) so that sectioning is possible. Solutions that demineralize tissue are composed of harsh chemicals that damage the soft tissues as well as the bone. Therefore, although demineralization may be necessary, it may also damage the morphology of the sample. Even so, the pathologist is usually able to complete histopathologic evaluation. The process of demineralization usually takes several days, depending on the amount of the bone and degree of mineralization. It may take 5 or 6 days to demineralize a large toe of a mature, large breed, male dog.

Tissue Processing and Histochemical Staining Techniques

After samples are trimmed and placed in labeled tissue cassettes (Figures 5-7 and 5-8), they are processed in a series of solutions designed to replace the water with paraffin wax. The processing is performed in automated machinery. The tissues are first washed in water to remove formalin, then dehydrated in graded ethanols to remove the water, and cleared in xylene or xylene substitute solutions to remove the ethanol, and are finally infiltrated and embedded in heated (liquid) paraffin wax (Figure 5-9). As the paraffin wax cools, it becomes firm and provides a matrix to the tissues so they can be cut in sections of 5 or 6 microns (Figure 5-10). Unless the tissues are firm (*e.g.*, in paraffin wax), it is not possible to obtain thin enough sections for precise microscopic examination. These 5 to 6 micron sections are placed on glass slides. The paraffin wax is melted slightly, and the samples are mechanically processed in another series of solutions to replace the paraffin wax with water so that the sections can be stained with water soluble dyes, also called histochemical stains. These dyes differentially stain cells and other tissue components facilitating microscopic evaluation. The hematoxylin and eosin (H&E) stain is the standard histochemical stain used in histology laboratories (Figure 5-11). The pathologist uses H&E stained sections to assess sample architecture and nuclear and cellular detail. Pathologists may request other stains to help differentiate similar looking cells (*e.g.*, mast cells versus histiocytic cells) (Figure 5-12) or to identify microorganisms (Figure 5-13). Table 5-1 provides examples of histochemical stains and the purpose of the stains. Additional time is required for performing and evaluating specially stained sections.

5-7 Photograph of tissue samples in a cassette.

5-8 Photograph of tissue cassette being closed.

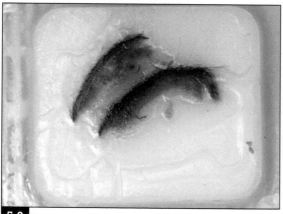

5-9 Photograph of tissue embedded in a paraffin block. These 'blocks' are saved for several years (time frame depends on the protocol used by the processing laboratory).

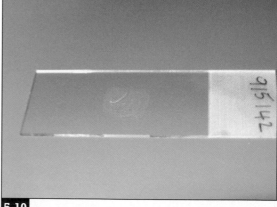

5-10 Photograph of a 5 micron section placed on the glass slide for staining.

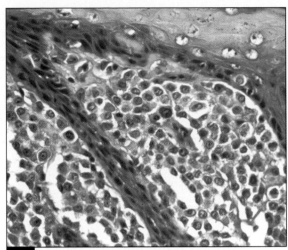

5-11 Photomicrograph of mast cell tumor mimicking cutaneous histiocytoma, (hematoxylin and eosin stain).

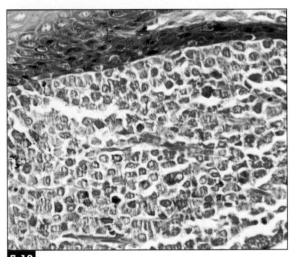

5-12 Photomicrograph of mast cell tumor depicted in Figure 5-11. Note magenta (metachromatic) intracellular granules (Giemsa stain).

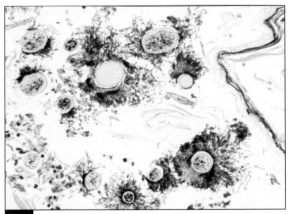

5-13 Photomicrograph of skin section. Note blue filamentous bacteria (Gram positive bacteria) oriented around hair shafts (Brown and Brenn stain).

Table 5-1 Selected Examples of Staining Procedures that Facilitate Histologic Evaluation of Cellular or Interstitial Constituents or Etiologic Agents

ELEMENTS BEING STAINED	STAIN	RESULTS
General purpose nuclear and cytoplasmic staining	Hematoxylin and eosin	Nuclei stain blue and cytoplasmic elements stain various shades of pink.
Cytoplasmic granules		
Iron pigments (hemosiderin)	Gomori's iron reaction Perl's potassium ferrocyanide	Iron stains blue.
Mast cell granules	Toluidine blue Giemsa	Mast cell granules stain magenta (metachromatic).
Melanin granules	Fontana-Masson silver method Warthin-Starry at pH 3.2	Melanin granules stain black (not specific for melanin granules).
	Melanin bleach methods remove melanin pigment. Examples include: • Potassium permanganate method • Performic or peracetic acid methods	Remove melanin pigment in heavily pigmented melanomas so that cytoplasmic and nuclear morphology are not obscured, and are better evaluated microscopically.
Cytoplasmic material		
Plasma cells	Methyl green pyronin	Plasma cell cytoplasm (RNA containing cytoplasm) stains red.
Interstitial materials		
Acid mucopolysaccharides	Alcian blue Colloidal iron	Acid mucopolysaccharides stain blue. Acid mucopolysaccharides are increased in myxedema and mucinosis.
Amyloid	Congo red examined with polarized light	Amyloid stains red and has green birefringence when viewed with polarized light.
Basement membrane	Periodic acid-Schiff (PAS)	Basement membrane stains purplish red.
Calcium	von Kossa	Mineral salts stain black.
Collagen	Masson's trichrome	Mature collagen stains blue.
Elastic fibers	Verhoeff-van Gieson	Elastic fibers stain blue-black to black.
Lipids: frozen sections on fresh or formalin-fixed tissue	Oil-red-O	Lipid stains orange red to brilliant red.
Muscle	Masson's trichrome	Muscle stains red.
	Phosphotungstic acid-hematoxylin (PTAH)	Muscle stains blue to purple.
Reticulum fibers	Wilder's reticulum	Reticulum fibers stain black.
Etiologic agents		
Bacteria		
Acid fast	Fite-Faraco modification (more sensitive) Ziehl-Neelsen, Kinyoun's methods	Acid fast bacteria (*Mycobacteria sp.*) stain red. *Nocardia* sp. may stain weakly acid fast (red).

continued

Table 5-1 Continued

ELEMENTS BEING STAINED	STAIN	RESULTS
Gram staining	Brown and Brenn MacCallum-Goodpasture	Gram positive bacteria stain blue. Gram negative bacteria stain red.
Spirochetes	Warthin-Starry	Organisms stain black.
Fungi	Periodic acid-Schiff (PAS)	Fungal elements stain purplish red with PAS.
	Gomori's methenamine- silver nitrate (GMS)	Fungal elements stain black with GMS.
	Mayer's mucicarmine	The capsule of *Cryptococcus sp.* stains bright carmine red with Mayer's mucicarmine.
Protozoa	Giemsa	*Rickettsia sp.* stain blue to violet *Leishmania sp.* stain reddish blue
Pythium species	Gomori's methenamine- silver nitrate (GMS)	The cell walls of *Pythium sp.* stain black.

Immunostaining Techniques (Direct Immunohistochemistry and Immunofluorescence) for Autoimmune Skin Disease

In the diagnostic setting, the use of direct immunostaining for immunoglobulin and/or complement detection has significantly declined as a diagnostic tool due to the high cost and lack of specificity and sensitivity of these staining techniques. Because of the high prevalence of false negative and false positive results with direct immunofluorescence and immunohistochemistry, if these tests are performed, they should always be evaluated in conjunction with clinical findings and conventional histopathology. Newer techniques that allow for more specific antigen targeting such as a desmosomal protein (desmoglein), use of "salt-split" skin sections for subepidermal bullous diseases, use of better substrates for indirect immunostaining, and use of immunoblotting and ELISA are facilitating diagnosis of immune mediated skin diseases. These techniques should greatly improve diagnostic accuracy for immune mediated skin disease in the future. Currently, samples for immunofluorescence evaluation are usually placed in Michel's medium, which preserves immunoglobulin and complement. However, some laboratories perform immunofluorescence on formalin fixed paraffin embedded tissue.

Immunohistochemistry (*e.g.*, immunoperoxidase) staining is performed on formalin fixed samples, as long as the samples are not in formalin longer than 48 hours. Prolonged fixation in formalin results in cross-linking of protein. General recommendations for collection of samples for immune mediated skin diseases are as listed in Table 5-2. It is also advisable to contact the laboratory personnel for specific recommendations as some tests may be discontinued and new and better tests offered in the future.

Table 5-2 Biopsy Sampling for Immune Mediated Skin Disease (Direct Immunofluorescence or Direct Immunohistochemistry)*

DISORDER	SAMPLES TO COLLECT
Pemphigus	Early primary lesions (e.g., papules, pustules). Nonlesional skin may or may not yield positive results.
Lupus erythematous	Older lesions from haired skin may be most reliable (in humans, biopsy samples from lesions over 1 month old are more likely to be positive than newer lesions).
Vesicular or bullous disease	Do not sample the vesicles or bullae, sample the skin adjacent to the vesicular/bullous lesions (either normal or erythematous skin).
Vasculitis	Lesions less than 24 hours old

PITFALLS	EXPLANATION
False positive and false negative results are common, thus these tests need to be interpreted with this knowledge in mind.	These techniques only allow detection of antibodies that target skin proteins, not disease-specific autoantigens or pathogenic versus nonpathogenic epitopes in an individual antigen. In addition, many technical factors prevent the detection of autoantibodies in immune mediated skin diseases, or identify autoantibodies in nonimmune mediated skin diseases or in normal animals.
Avoid lesions that may be secondarily infected.	Increases likelihood of false positive result (immunoglobulin or complement deposition may be the result rather than the cause of the skin disease).
Avoid sampling if animal has been on glucocorticoid or other immunosuppressive therapy.	Increases likelihood of a false negative result (reduces immune response and presence of antibody or complement in tissue).
Avoid collecting nasal planum and footpads for immunostaining.	Increases likelihood of false positive result (normal dogs and normal cats may have positive staining of nasal planum or footpad tissue).
Avoid prolonged formalin fixation for immunostaining (time in formalin fixative should be 48 hours or less).	Prolonged fixation in formalin can result in cross linking of protein and increase false negative result.

FIXATION	
Immunofluorescence	Michel's media**
Immunohistochemistry (e.g., immunoperoxidase)	10% neutral buffered formalin for 48 hours or less

*Remember, immunostaining is not highly sensitive or specific, and false positive and false negative results are common. Because of the lack of sensitivity and specificity, results of these techniques should always be interpreted in conjunction with the clinical and histologic findings. New techniques currently being developed will replace direct immunofluorescence and immunohisto-chemistry, and they will greatly improve diagnostic accuracy of immune mediated skin disease in the future.

**Immunofluorescence also can be performed on formalin fixed samples; however, direct immunofluorescence is in declining use. Check with the laboratory personnel performing the immunologic tests for the preferred method of sample preservation and shipment.

Immunohistochemical Staining Techniques for Tumor and Tumor-like Conditions

Immunohistochemical staining can be a useful diagnostic tool for aiding in the differentiation of tumors or tumor-like conditions when such differentiation is not possible using conventional histopathologic evaluation (Table 5-3). Immunohistochemical staining procedures use the specificity of one or a group of antibodies directed against cell surface or cytoplasmic proteins to identify cell type regardless of histologic morphology (Figure 5-14a and 5-14b). For example, keratin filaments in epithelial cells are not typically present in mesenchymal cells so can be used to differentiate a spindle cell squamous cell carcinoma from a fibrosarcoma. However, none of the antibodies is perfectly specific for their antigen, so interpretation requires careful consideration of clinical and other histopathologic information. Also, for some immunohistochemical procedures, fresh or frozen specimens are favored over formalin fixed specimens. Evaluation of a series (panel) of antibodies is preferred, as the pattern of staining with a panel of antibodies is more reliable than staining with one or two antibodies. It is best to check with the personnel at the laboratory performing the tests for the preferred method of sample preservation, and for the tumors that can be identified with the antibody panels they are using.

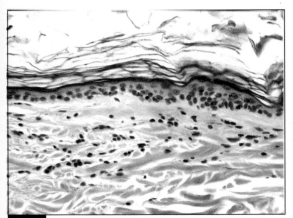

5-14a Photomicrograph of H&E stained skin sample in a dog with epitheliotropic lymphoma.

5-14b Photomicrograph of serial section of skin in Figure 5-14a, stained for CD8+ antigen (T-lymphocyte antigen). Note: the CD8+ cells have red cytoplasm. The extent of epidermal infiltration could easily be missed on the H&E sections.

Table 5-3 Example of Immunohistochemical Differentiation of Lymphocytic and Histiocytic Tumors or Tumor-like Conditions

TUMOR OR CONDITION	ANTIGEN (MARKER)					
	CD1*	CD4	MHC II	ICAM 1	CD90 (thy-1)	CD3
Cutaneous histiocytoma	+	-	+	+	-	-
Canine reactive histiocytosis	+	+	+	+	+	-
Canine disseminated histiocytic sarcoma (malignant histiocytosis)	+	-	+	+	+/-	-
Epitheliotropic lymphoma	-	Most - few +	Most - few +	Most -	-	+

CD–cluster of differentiation; MHC–major histocompatibility complex; ICAM–intercellular adhesion molecule
*CD1–can require use of fresh tissue

Immunohistochemical Staining Techniques for Infectious Agents

Infectious agents are an important cause of skin diseases. Therefore, documentation of an infectious agent in the tissue can help achieve a definitive diagnosis. Some agents are readily visible in H&E stained sections. However, the identification of other infectious agents requires use of special histochemical staining procedures that are listed in Table 5-1. Even so, some infectious agents are not visible histologically even if they are the cause of the lesions. For example, feline herpesvirus infection may cause facial dermatitis with eosinophils mimicking a hypersensitivity reaction or an eosinophilic granuloma, but inclusion bodies may not be present. Specific identification of herpesvirus in lesional tissue via immunohistochemistry helps achieve a definitive diagnosis. Other infectious agents that can be identified by immunohistochemical staining are papillomavirus, feline leukemia virus, canine distemper, *Leishmania donovani*, *Sarcocystis canis*, and *Toxoplasma gondii*. Immunofluorescence techniques are available for some infectious agents, particularly the systemic fungal infections. These staining techniques provide specific identification of the organism present in the tissue. Personnel at the laboratories performing these techniques can provide information on the best methods of sample preservation and shipment.

Other Techniques

Other types of fixation, processing, or sectioning can be performed in selected cases. The most common of these is the technique of frozen sectioning. Frozen sections can be prepared in minutes to hours versus days for standard processing techniques, and do not use the chemical solutions required by standard processing techniques. The tissue is hardened for sectioning by freezing rather than being embedded in paraffin wax. The sample is mounted in tissue embedding media, and quickly frozen in liquid nitrogen or with carbon dioxide. The sample is frozen so quickly that ice crystals do not form, and thus the sample architecture is sufficiently preserved for microscopic examination (in contrast to slow freezing that damages tissue by formation of ice crystals). Therefore, frozen sectioning is used either when rapid evaluation of the tissue is necessary (to determine if a tumor is benign or malignant before performing extensive surgery), or when evaluating for a substance, such as lipid, removed by standard processing techniques. When using frozen sectioning to diagnose tumors while a patient is undergoing surgery, close physical access to the processing laboratory is necessary thus the procedure is generally limited to veterinary university teaching hospitals. Frozen sectioning requires use of a special laboratory instrument called a cryostat.

There are other specialized fixation, processing, or staining techniques including those for tissues examined via electron microscopy, for selected tissues such as non-demineralized bone, and for specialized test procedures such as polymerase chain reaction and in situ hybridization. The discussion of fixation, processing, and staining for these specialized techniques is beyond the scope of this book.

Section 6

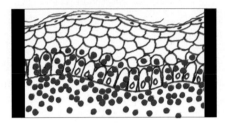

Glossary

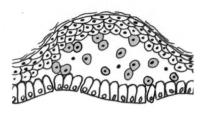

Acantholysis

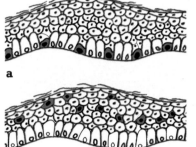

a

b

Apoptosis

A

Acantholysis: loss of cohesion between keratinocytes due to the breakdown of intercellular attachments as a result of immune destruction or other processes such as neutrophilic enzymatic destruction.

Acanthosis: thickening of the spinous cell layer (stratum spinosum)

Acral: distal parts of the extremities

Alopecia: complete hair loss

Anagen: phase of hair cycle in which hair growth (synthesis) takes place

Anaplasia: lack of cellular differentiation and organization, a feature of neoplastic cells

Apoptosis: physiological or programmed cell death. Apoptosis may affect basal cells (a) or keratinocytes (b)

Atrophy: reduction in size of a cell, tissue, organ, or part

B

Blister (vesicle or bulla): localized collection of fluid usually in or beneath epidermis

Bulla: large blister (> 1.0 cm)

C

Callus: a thick, hard, hairless plaque usually with accentuation of superficial skin architecture (creases)

Catagen: transition phase of the hair cycle between growth and resting phases

Comedo (comedones): plug of cornified cells and sebum in a dilated hair follicle

Cornification: production of stratum corneum, which includes keratin, lipids, and other components of this layer

Crust: material formed by drying of exudate or secretion on the skin surface

D

Dermatosis: non-specific term to denote any cutaneous abnormality or eruption

Dermatophytosis: infection of the cornified cells of the epidermis, hair, or claws with fungi of the genera *Microsporum*, *Epidermophyton*, or *Trichophyton*

Dyskeratosis: abnormal, premature, or imperfect keratinization

Dysplasia: abnormal development

E

Effluvium: shedding of hair

Epidermal collarette: a ring of scale that expands peripherally

Epidermolysis: lysis (dissolution) of a portion of the epidermis

Epitheliotropic lymphoma: T-cell lymphoma targeting the epidermis and adnexal epithelium

Eruption: rapid development of skin lesion associated with redness or prominence or both

Erosion: superficial loss of a portion of the epidermis, which leaves the basement membrane intact

Erythema: redness of skin due to congestion of capillaries

Excoriation: superficial loss of epithelium due to physical trauma (scratching)

Exfoliation: shedding of layers or scales

Exudation: escape of fluid, cells, or debris from blood vessels and its deposition in or on other tissues

F

Fissure: a linear defect (cleft or groove) within the epidermis or dermis

Folliculitis: inflammation of a hair follicle

Furuncle: circumscribed, frequently painful nodule (accumulation of pus) in the dermis secondary to follicular rupture

G

Glabrous: smooth skin, hairless skin

H

Hamartoma: a disturbance of tissue growth in which the growth of cells of a circumscribed area surpasses the growth of cells of the surrounding area, and presumed to be of congenital origin

Hydropic degeneration: intracellular edema of basal epidermal cells

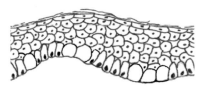

Hydropic degeneration

Hyperhidrosis: excessive sweating

Hyperkeratosis: histologic term for thickening of stratum corneum (cornified layer)

Hyperpigmentation: an increase in epidermal and/or dermal melanin pigment

Hyperplasia: increase in the number of normal cells

Hypopigmentation: decrease in epidermal and/or dermal melanin pigment

Hypoplasia: incomplete development

Hypotrichosis: less hair than normal (partial loss of hair)

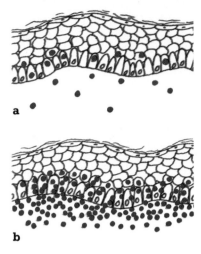

Interface inflammation

I

Ichthyosis: congenital skin disorder depicted by marked hyperkeratosis

Impetigo: bacterial dermatitis characterized by epidermal pustules

Indurated: hardened

Interface inflammation: inflammation that targets the dermal epidermal junction (interface) with damage to the basal cell layer (hydropic degeneration and/or apoptosis). The dermal infiltrate is mononuclear and may obscure the dermal epidermal junction. The inflammation may be referred to as cell poor (a) or cell rich (b) depending on the amount of cellular infiltrate.

Intertrigo: dermatitis that develops from friction between opposing skin surfaces, *e.g.*, adjacent skin folds

K

Keratinocytes: the epidermal cells that synthesize keratin and comprise over 90% of epidermal cells

Keratosis: elevated circumscribed area of excessive production of stratum corneum

L

Langerhans cells: antigen presenting cells of the skin

Lichenification: thickening of skin with accentuation of skin architecture (creases)

Lichenoid: dense "band-like" dermal inflammation parallel to the epidermis that may or may not obscure the epidermal dermal interface. There is controversy regarding the meaning of the term "lichenoid". Some pathologists define lichenoid as a dense band of subepidermal inflammation that is not associated with degeneration of basal cells and that does not obscure the epidermal dermal interface. In contrast, other pathologists define lichenoid as a dense band of subepidermal inflammation that is associated with basal cell degeneration and that can obscure the epidermal dermal interface (*e.g.*, consistent with cell rich interface dermatitis).

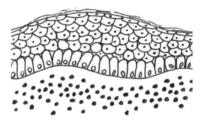

Lichenoid inflammation

M

Macule: flat circumscribed lesion of altered skin color (< 1 cm)

Melanin: dark, amorphous pigment consisting of dihydroxy indoxylic acid

Melanophage: macrophage containing ingested melanin

Merkel cell: a neuroendocrine cell found in the stratum basale

Microabscess, Munro: small collection of desiccated neutrophils within or below the stratum corneum

Microabscess, Pautrier: small collection of neoplastic lymphocytes within the epidermis

Mycelium: a mass of hyphae

Mycosis fungoides: synonym for epitheliotropic lymphoma. The term was used initially in 1806 to describe lesions in a person in which the skin rash evolved into tumors resembling mushrooms.

Myxedema: thickening of the skin due to abnormal deposits of dermal mucin

Microabscess, Munro

Microabscess, Pautrier

N

Nevus: circumscribed malformation of the skin assumed to be of congenital or inherited origin, and consisting of any component of the skin. The term "hamartoma" is preferred to avoid confusion with the pigmented nevi (moles) in human dermatology and dermatopathology.

Nodule: a circumscribed, solid elevation of the skin (> 1 cm)

O

Onychodystrophy: abnormal formation of the claw

Onychomadesis: sloughing of claws

P

Panniculitis: inflammation of subcutaneous adipose tissue

Papule: small, circumscribed, solid elevation of skin (< 1 cm)

Parakeratosis: retention of pyknotic nuclei in epidermal cells of the stratum corneum due to faulty or accelerated cornification

Paronychia: inflammation of skin around the claws

Pemphigus: vesicular to pustular autoimmune disease of the skin and, in some cases, the mucous membranes in which the pathomechanism involves acantholysis.

Phaeohyphomycosis: mycotic disease caused by pigmented fungi of a variety of genera and species that do not form sclerotic bodies or granules

Pigmentary incontinence: the process whereby melanin pigment released from injured basal layer cells is phagocytized by macrophages in the superficial dermis or may remain free in the dermis

Plaque: a flat topped, solid elevation in the skin that occupies a relatively large surface area in comparison with its height

Pruritus: itching

Pyoderma: pyogenic (pus producing) bacterial infection of the skin, *e.g.*, luminal folliculitis

Pustule (non follicular): small, circumscribed accumulation of pus within the epidermis

S

Salt-split technique: a technique used in immunofluorescence studies of the skin where sodium chloride is used to split the epidermis from the dermis though the lamina lucida. The technique allows better differentiation of some immune mediated disorders such as bullous pemphigoid, linear IgA dermatosis, and epidermolysis bullosa acquisita.

Scale: a fine or thin and plate-like loose fragment of stratum corneum resting on the surface of skin

Scar: an area of fibrous tissue that has replaced components of the epidermis, dermis, and/or subcutis

Sebaceous: pertaining to sebum

Seborrhea: nonspecific term for clinical signs of scaling, crusting, and greasiness. Primary seborrhea is a more specific term applied to inherited cornification disorders.

Sebum: secretion of sebaceous glands; thick semifluid substance composed of fat and epithelial debris

Spongiosis: intercellular edema that, by widening of the intercellular spaces and stretching of the "intercellular junctions", creates a sponge-like appearance to the epidermis.

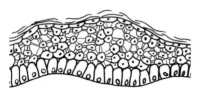

Spongiosis

T

Telogen: resting phase of the hair cycle

Tumor: an abnormal mass of tissue, the growth of which exceeds and is uncoordinated with that of normal tissue and persists in the same excessive manner after cessation of the stimuli which evoked the change (Willis, 1948)

U

Ulcer: loss of epidermis and the basement membrane

Urticaria: transient, raised edematous often erythematous lesion (hive) due to increased vascular permeability

V

Vesicle: localized collection of clear fluid usually within or below the epidermis (< 1.0 cm). Vescles can be subcorneal (a), suprabasilar (b) or subepidermal (c).

Vibrissa: (sinus hair, whisker) long, coarse hair located about the muzzle and containing a vascular sinus and numerous nerve fibers

Vitiligo: acquired disorder characterized by circumscribed areas of depigmentation in the skin

W

Wheal: a transient, smooth, circumscribed, slightly elevated area on skin caused by dermal edema

Y

Yeast: unicellular budding fungus

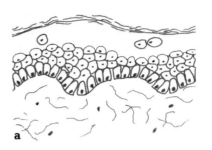

a

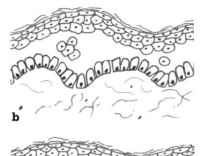

b

c

Vesicles

Section 7

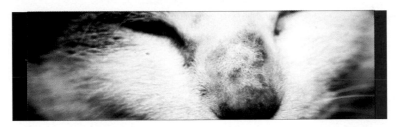

Differential Diagnoses of Clinical Lesions and Lesions in Selected Anatomic Locations

7-1 Alopecia due to congenital hypotrichosis, 12 week old Rottweiler. (Courtesy of Dr. Ralf Mueller.)

7-2 Alopecia due to follicular dysplasia, ventral neck, Curly coated retriever. (Courtesy of Dr. Ralf Mueller.)

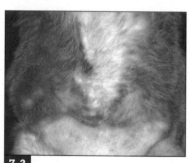

7-3 Hypotrichosis associated with atopic dermatitis, ventral abdomen, Abyssinian cat.

This section is provided to help practitioners formulate clinical differential diagnoses based on clinical lesions and anatomic distribution patterns. Specifically, it is hoped that these lists will: 1) help the practitioner provide differential diagnoses on pathology submission forms, 2) better visualize how diverse clinical conditions may have similar clinical lesions, 3) through cross-reference with Sections 1 and 2, learn which biopsy sampling techniques are most useful for various lesions, and 4) through cross-reference with Section 3, gain a better understanding of the disease process that is the basis of the clinical lesions.

Clinical Lesions

(blue denotes more common causes)
(* denotes uncommon causes)

Alopecia Canine Figures 7-1 and 7-2

Infectious agents (e.g. bacteria, dermatophytes, Demodex sp., yeast)..................................... 99

Self induced (e.g. hypersensitivity reactions, psychogenic, parasitic, boredom).. 93

Follicular dysplasia (e.g. color dilution alopecia, pattern alopecia, idiopathic or cyclic flank alopecia syndrome) 95

Congenital/hereditary alopecia (localized or generalized e.g. Mexican hairless breed)................................... 95

Endocrine (e.g. hypothyroidism, hyperadrenocorticism, hyperestrogenism) 95

Alopecia X of plush coated breeds (a group of disorders not a single cause).. 95

Other

Dermatomyositis 85

Vaccine-induced ischemic dermatitis...................... 85

Post-clipping alopecia................................... 95

Anagen defluxion and telogen effluvium 95

Sebaceous adenitis (scaling typical, but may be subtle)....... 101

Glucocorticoid therapy (localized or generalized) 95

Traction alopecia*...................................... 95

Mural folliculitis and alopecia areata* 98, 90

Alopecia Feline Figure 7-3

Self-induced (e.g. hypersensitivity: atopy, food; psychogenic) 93

Infectious agents (e.g. dermatophytes, fleas, Demodex sp., other parasites)... 99

Congenital/hereditary alopecia (localized or generalized e.g. Canadian hairless {Sphinx} breed) 95

Feline mural folliculitis* 94

Feline paraneoplastic alopecia* 95

Anagen defluxion/telogen effluvium*........................ 95

Crusts Figures 7-4 and 7-5

Erosive and Ulcerative Dermatoses Figure 7-6 and 7-7

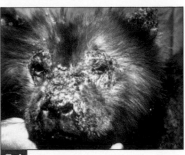

7-4 Marked scale-crust due to pemphigus foliaceus, Chow chow. (Case material Dr. Alexander Werner.)

7-5 Adherent crusting due to self trauma, atopic cat.

7-6 Erosions, crusts and significant ulceration due to mosquito bite hypersensitivity, domestic short hair cat.

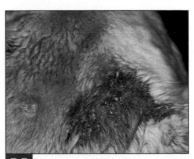

7-7 Pyotraumatic dermatitis (hot spot) due to self trauma associated with atopic dermatitis, dog.

continued

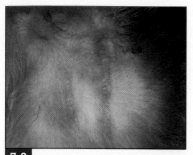

7-8 Feline linear granuloma associated with atopic dermatitis, caudal thigh, cat depicted in Fig 7-3.

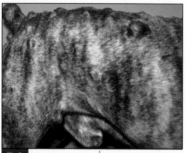

7-9 Follicular cysts, trunk, Boxer. (Courtesy Teton NewMedia, *Dermatology for the Small Animal Practicitioner.*)

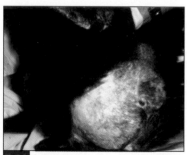

7-10 Mast cell tumor, shoulder, cat. (Courtesy of Dr. Ralf Mueller.)

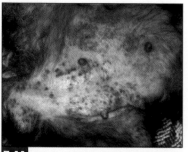

7-11 Papules, crusted papules, and macules associated with atopic dermatitis, ventral abdomen, poodle.

Feline Plaques or Ulcers Figure 7-8

Nodules Figures 7-9 and 7-10

Nodules unique to or often seen in the cat

Papules Figure 7-11

Feline multifocal and crusted papules (feline miliary dermatitis) Figure 7-12

Pruritus Figures 7-13 and 7-14

Pustules Figure 7-15

Note pustules are extremely uncommon in the cat, we tend to see the ruptured sequelae (e.g. crusts)

continued

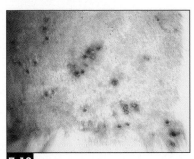
7-12 Miliary dermatitis due to hypersensitivity to flea bites, cat.

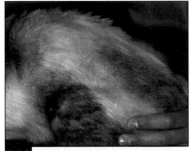
7-13 Erythema, alopecia and hyperpigmentation due to over grooming associated with hypersensitivity to flea bites, Siamese cat.

7-14 Pruritus due to chronic generalized pemphigus foliaceus, Spitz.

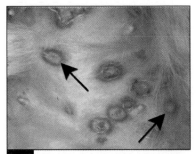
7-15 Large, flaccid pustules (arrows) due to pemphigus foliaceus, ventral abdomen, dog. Epidermal collarettes are also present. (Courtesy of Dr. Ralf Mueller.)

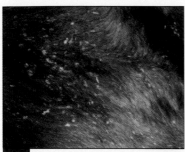

7-16 Seborrhea sicca due to hypothyroidism, Labrador retriever. (Courtesy of Dr. Ralf Mueller.)

7-17 Vesicles (arrows) due to bullous pemphigoid, inner pinna, dog. (Courtesy of Dr. Rod Rosychuk.)

7-18 Hypotrichosis, erythema and punctuate ulcers due to mosquito bite hypersensitivity, nasal planum and dorsal muzzle, cat.

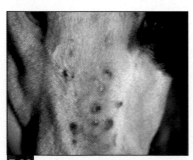

7-19 Papules, hypotrichosis, and hemorrhagic bullae due to actinic dermatitis, dorsal muzzle, dog. (Courtesy of Dr. Ralf Mueller.)

Scaling Figure 7-16

Vesicles Figure 7-17

Regional dermatoses
Facial dermatitis Figures 7-18 and 7-19

Hypersensitivity (atopy, food

Hypersensitivity (mosquito bites in cats)

Feline eosinophilic granuloma complex

Canine eosinophilic folliculitis and furunculosis (arthropod bite)*

Infections (e.g. staphylococcal, *Demodex* sp., dermatophytes, *Cryptococcus* sp., opportunistic fungi)

Cutaneous (discoid) lupus erythematosus

Pemphigus foliaceus and erythematosus

Solar dermatitis

Neoplasia, especially squamous cell carcinoma

Drug eruption*

Uveodermatological syndrome (synonym: Vogt-Koyanagi-Harada-like syndrome)*

Vitiligo*

Feline herpesvirus dermatitis*

Facial dermatitis of Persian cats*

Footpad dermatitis Figures 7-20 and 7-21

Pemphigus foliaceus

Zinc responsive dermatosis

Superficial necrolytic dermatitis

Feline plasmacytic pododermatitis

Cutaneous (discoid) lupus erythematosus

Idiopathic nasodigital hyperkeratosis*

Acrodermatitis of Bull terriers*

Canine distemper*

Familial digital hyperkeratosis*

Cutaneous horn*

Pinnal dermatoses Figures 7-22 and 7-23

Fly bites

Sarcoptes scabiei

Bacterial and fungal/yeast infection of the pinnae

Ear margin seborrhea

Vasculitis

Solar dermatitis in white cats

Feline pinnal alopecia

Pemphigus foliaceus (especially inner pinna)

Pattern alopecia

Dermatomyositis

Bullous pemphigoid (especially inner pinna)*

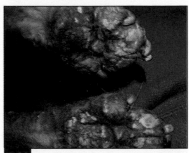

7-20 Fissured, hyperkeratotic footpads due to acrodermatitis, Bull terrier. (Courtesy of Dr. Ken Mason.)

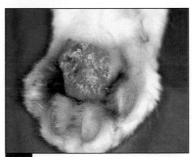

7-21 Scale-crust, lichenification, and spongiotic swelling due to plasmacytic pododermatitis, metacarpal footpad, cat.

7-22 Ulcer due to vasculitis, ear margin, dachshund. (Courtesy of Dr. Wayne Rosenkrantz.)

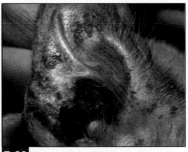

7-23 Focal scale-crust due to pemphigus foliaceus, inner pinna, cat. (Courtesy of Dr. Ralf Mueller.)

continued

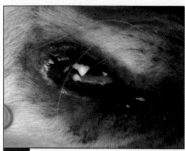

7-24 Alopecia, erythema, erosions, depigmentation, and crusts of lip margins (mucocutaneous pyoderma) associated atopic dermatitis, Golden retriever.

7-25 Alopecia, scale-crust, and ulceration due to dermatomyositis, tip of the tail, Chow chow. (Courtesy of Dr. Ralf Mueller.)

Mucocutaneous junctions Figure 7-24

Mucocutaneous pyoderma

Intertriginous pyoderma (especially lip fold and vulvar fold)

Cutaneous (discoid) lupus erythematosus

Superficial necrolytic dermatitis (synonym: metabolic epidermal necrosis)*

Pemphigus vulgaris and bullous pemphigoid*

Mucocutaneous candidiasis*

Peripheral anatomic locations (ear tips, tail tip) Figure 7-25

Dermatomyositis

Other forms of vasculitis (e.g. systemic disease, drug or vaccine reaction)

Cryoglobulin disorder*

Frost bite*

Appendix A

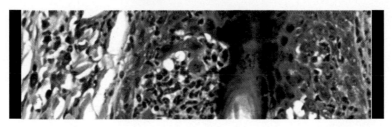

Dermatopathology Submission Form

Dermatopathology Submission Form

Dr._____   Date of this submission _____

Hospital _____   Tests requested histopathology: _____

Street address_____   other:_____

City, State, Zip code _____

Telephone_____Facsimile_____   **Current chief complaint** (use back if necessary):

Animal name (first and last):   _____

Species _____Breed _____Age _____Sex_____   _____

Clinical History:

Date the current condition began _____

Lesion appearance early in disease_____

Systemic illness   ☐ no ☐ yes explain _____

Previous skin or ear problems   ☐ no ☐ yes explain _____

Other animals or people affected ☐ no ☐ yes explain _____

Lesion distribution and type:

Mark biopsy site with "x"   Pruritus ☐ no ☐ yes   Seasonal ☐ no ☐ yes

Shade lesion location   Severity (circle)   mild 1 2 3 4 5 6 7 8 9 10 severe

Location:

☐ Face ☐ Feet ☐ Rump ☐ Abdomen ☐ Neck

Symmetry ☐ no ☐ yes   ☐ Ears ☐ Legs ☐ Tail ☐ Flank ☐ Other _____

Please check those lesions present:

☐ alopecia   ☐ lichenification
☐ callus   ☐ macule
☐ comedo   ☐ claw lesions
☐ crust   ☐ nodule
☐ depigmentation   ☐ papule
☐ epidermal   ☐ plaque
  collarette   ☐ pustule
☐ erosion   ☐ scale
☐ erythema   ☐ scar
☐ excoriation   ☐ ulcer
☐ footpad lesions   ☐ vesicle
☐ hyperpigmentation   ☐ other _____

Previous diagnostic testing (if yes, please list and/or attach results):

Skin scrapings	☐ no ☐ yes _____		Allergy testing	☐ no ☐ yes _____	
Surface cytology	☐ no ☐ yes _____		CBC, Chemistry	☐ no ☐ yes _____	
Bacterial C & S	☐ no ☐ yes _____		Hormone assays	☐ no ☐ yes _____	
Fungal culture	☐ no ☐ yes _____		Immunology (ANA)	☐ no ☐ yes _____	
Elimination diet	☐ no ☐ yes _____		Biopsy	☐ no ☐ yes _____	
Wood's light/hair	☐ no ☐ yes _____		Other	_____	

Previous Treatment: **If yes, please explain:**

Antibiotic*	☐ no ☐ yes	Type _____	Duration _____	Response ___%
Antihistamine	☐ no ☐ yes	Type _____	Duration _____	Response ___%
Anti-yeast/fungal	☐ no ☐ yes	Type _____	Duration _____	Response ___%
Glucocorticoid	☐ no ☐ yes	Type _____	Duration _____	Response ___%
Shampoo therapy	☐ no ☐ yes	Type _____	Duration _____	Response ___%
Flea control	☐ no ☐ yes	Type _____	Duration _____	Response ___%
Anti-scabies	☐ no ☐ yes	Type _____	Duration _____	Response ___%

Other _____

*Antibiotic: Did lesions resolve, but recur when therapy was stopped, or did lesions not respond at all? _____

List clinical differential diagnoses (VERY IMPORTANT!): _____

_____

Appendix B

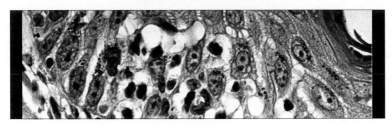

Suppliers

Suppliers

Formalin

Many companies supply formalin. Supplies may be available from the laboratory or pathologist processing your biopsy samples. Any liquid that is shipped via airplane has the potential to leak from bottles (due to pressure changes associated with altitude). Some commercial suppliers of formalin have designed their bottles and lids to minimize this problem. Discussion with the suppliers regarding this problem is recommended. The following supplier has been used by one of the authors with good results:

Richard-Allan Scientific

Kalamazoo, MI 49007

800-522-7270

Michel's Media

Zeus Scientific, Inc

Raritan, N.J. 08869

800-286-2111

Biopsy Punch Instruments

Acuderm biopsy punch (Acu-Punch)

Acuderm, Inc.

Ft. Lauderdale, FL 33309

800-327-0015

Dyes for Color Coding Surgical Margins or for Marking Specimens

Bradley Products, Inc. (The Davidson Marking System)

Bloomington, MN   55420

800-325-7785

Tissue Cassettes for Labeling Specimens

Richard-Allan Scientific

Kalamazoo, MI 49007

800-522-7270

Appendix C

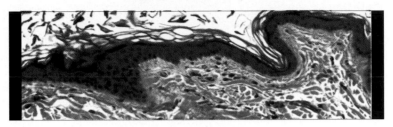

Case Review

Case Review

Collection of biopsy samples does not replace the necessity of the dermatologic examination, but serves as one component of the diagnostic process. Evaluation of skin biopsy samples is not a substitute for evaluation of skin scrapings, tape impressions, skin surface cytology, or fungal culture etc. In fact these dermatologic procedures can help guide the biopsy selection process. Thought must be given to recent medications that can influence inflammation and other aspects of skin lesion morphology, particularly glucocorticoids or other anti-inflammatory agents. Depending on the health of the patient, it is usually prudent to wait to collect biopsy samples until the cutaneous effects of those drugs are eliminated, and it may be useful to institute antibiotic therapy to eliminate secondary bacterial infection prior to collecting skin biopsy samples. The two scenarios below are provided as general examples of the role played by dermatology and dermatopathology in case assessment. This process includes evaluating the animal for clinical lesions, making decisions about the best samples to collect, using appropriate techniques for collecting those samples, completing the dermatopathology submission form including a list of differential diagnoses, and interpreting the skin biopsy report. The case reviews will refer to other sections in this book, illustrating how this book can be used to assist the practitioner in diagnosing skin diseases of dogs and cats.

CASE 1:
How to Manage the Biopsy Sampling Process of a Chronically Itchy Animal

History and signalment:

✔Obtain a thorough history.

Summary of results: This patient is a ten-year old, spayed female, terrier cross-bred dog with a long history of allergic dermatitis and secondary pyoderma. She presents yet again with an owner complaint of red itchy skin, but more severe than in the past. The owner indicates that the degree of redness is unusually severe and more generalized. The dog appears to have some pain in the mouth. The patient has been a client at this clinic for the past 9 years.

Dermatologic examination:

✔Carefully evaluate the dog from head to toe for primary and secondary skin lesions. Note lesions in the eyes, oral cavity, lymph nodes, etc.

Summary of results: The cutaneous lesions consist of mild diffuse hypotrichosis with marked erythema and minute papules, several scattered small crusts, several epidermal collarettes on the abdomen, a few excoriations on trunk, and hyperpigmentation in inguinal and abdominal areas. No mites are identified via tape impressions or skin scrapings. An occasional *Malassezia* is identified in impression smears of axillary skin, but yeast occasionally were present in the past. The oral mucous membranes (gingiva, tongue, palate, and lips) are also reddened (not noted in past episodes).

Biopsy procedure:

✔First, determine if biopsy sampling is appropriate in this case *(Refer to Section 1 {including Table 1-1} for discussion of the situations in which biopsy sampling is essential or may be useful)*. In this case the owner is concerned that the lesions are worse than in past episodes, and that the skin is more intensely red. The dermatologic examination did not reveal an explanation for the increased lesion severity. However, the new oral lesions suggest there may be something different with this episode.

Summary of results: Ultimately, the decision is made to collect biopsy samples either to: a) confirm the clinical impression of a more severe episode of allergic dermatitis and secondary pyoderma, or 2) determine if there might be another problem contributing to the lesions that was not identified with the dermatologic examination.

☞ The decision is made to do the biopsy procedure before instituting treatment that could alter lesions.

✔Second, collect representative lesions. *(Refer to Section 2 for the biopsy techniques appropriate for each clinical lesion and Section 1 for the surgical techniques for those biopsy procedures).*

Summary of results: Four skin biopsy samples and one biopsy sample of the oral mucosa of lip are collected and placed in one bottle of 10% buffered formalin. The sample of oral mucosa is wrapped in lens paper to keep it separate from the skin samples. The samples consist of: 1) two 6 mm punch biopsy samples of papular, erythematous, hypotrichotic skin lesions, 2) two elliptical incisional samples (attached to cardboard) of epidermal collarettes collected from lesional through and including normal skin, and 3) one punch biopsy sample of erythematous oral mucosa from lip (wrapped in lens paper). The bottle of formalin is labeled with the animal and hospital name. A dermatopathology submission form (similarly labeled) is completed and describes the location of the lesions, locations of samples (oral sample in lens paper), types of lesions present, negative results of dermatologic examination, and a list of clinical differential diagnoses.

Using the biopsy report:

✔Read and interpret the biopsy report. *(Refer to Section 4 on tumors. The glossary is also available to help clarify new or unfamiliar terminology).*

Report Diagnosis: Lymphocytic infiltrates with epitheliotropism and Pautrier's microabscesses, skin and oral mucosa.

Report Comment: The lesions are compatible with epitheliotropic lymphoma (mycosis fungoides).

Summary of results: The diagnosis and comment indicate that the skin and oral lesions in this dog are neoplastic. The lesions consist of lymphoma targeting the epithelium.

Comments about this case:

From the history it was suspected that the dog had another episode of allergic skin disease and secondary bacterial infection. The allergy and pyoderma scenario is a much more likely possibility (allergy and pyoderma are much more common than epitheliotropic lymphoma). If the dog had allergy and pyoderma, the pathology report would have so indicated. In addition, presence or absence of underlying or contributing factors such as *Malassezia* or *Demodex* infestation would have been recognized. Even if allergic dermatitis had been documented, it is often not possible to identify the specific type of allergy by histopathologic evaluation alone, **limiting** the usefulness of biopsy sampling in these cases. Therefore, the biopsy sample evaluation in the allergic dermatitis complicated by pyoderma scenario would not have provided any more information than the dermatology examination provided, **except** that no other disease process was identified.

It can be difficult to determine when to and when not to use biopsy sampling as a component of the diagnostic process. In this case there were clues that biopsy evaluation was indicated: 1) the increased lesion severity (or a change in character of disease), 2) a clinical impression that there was more to the lesions than the dermatologic examination indicated, and 3) the presence of new oral lesions.

Dermatopathologic examination can confirm the clinical diagnoses and it can rule in or out other disease syndromes that potentially could be confused with those conditions. The practitioner and owner should have realistic expectations of what type of information a biopsy evaluation can and can not provide.

If a practitioner has any concerns about the validity of the diagnosis, the pathologist reading the case should be contacted. There may be other factors that influence interpretation of a report, and very rarely there may be laboratory errors. The pathologist who evaluated the case is the first person with whom to consult.

CASE 2:
How to Manage the Biopsy Sampling Process of the Animal with Prominent Scaling and Variable Alopecia

History and signalment:

✔Obtain a thorough history.

Summary of results: A new client to your hospital presents his four-year old, spayed female, standard poodle for evaluation. She has been losing hair and having "dandruff" for 6 months.

Dermatologic examination:

✔Carefully evaluate the dog from head to toe for primary and secondary skin lesions.

Summary of results: This dog has moderate hypotrichosis and patchy alopecia with scaling on dorsal trunk, face, and ears. The hair easily epilates. The dog appears otherwise healthy (active, energetic, atrophic mammary glands, and normal external genitalia). No demodectic mites or other organisms are seen in skin scrapings. No organisms are seen in tape impressions. DTM is performed, but no dermatophytes are seen in microscopic examination of hair shafts. There is no pruritus.

Biopsy procedure:

✔First, determine if biopsy sampling is appropriate in this case. *(Refer to Section 1 including Table 1-1 for discussion of the situations in which biopsy sampling is essential or may be useful).* The poodle is young and otherwise healthy. There is no evidence of allergy or infection as a source for the lesions. Dermatologic examination did not reveal a cause for the lesions, and there are multiple differential diagnoses (see below). Clinical pathologic evaluations for endocrine disease could be considered as an alternative to biopsy sampling, but the poodle appears healthy and active with no significant suggestion of endocrine anomaly.

Summary of results: Ultimately, the decision is made to collect biopsy samples to determine the source of the hypotrichosis and scaling.

✔Second, collect representative lesions. *(Refer to Section 2 for the biopsy technique appropriate for each clinical lesion and section 1 for the surgical techniques for those biopsy procedures).*

Summary of results: Three, 8 mm punch biopsy samples from alopecic scaling areas are collected and placed in a bottle of 10% buffered formalin. The bottle is labeled with animal and hospital name. A dermatopathology submission form (similarly labeled) is completed and describes the location of the lesions, types of lesions present, negative results of dermatologic examination, and a list of clinical differential diagnoses.

Section 2 provides information regarding differential diagnoses of conditions associated with alopecia, hypotrichosis, and scaling. Section 3 also provides differential diagnoses for lesions associated with hyperkeratosis (scaling) and follicular atrophy. These sections can help with the formulation of a list of differential diagnoses, and help make a determination regarding the value of biopsy sampling or clinical pathologic testing. In this case, biopsy sampling is a good choice.

Clinical differential diagnoses include: cornification or keratinization disorders, sebaceous adenitis, follicular infection, self trauma, follicular dysplasia, and endocrine disease. Follicular infection and self trauma are not likely (no mites seen in skin scrapings and no evidence of infection on dermatologic examination, and no history of pruritus). The diagnostic considerations are clinical pathologic testing for endocrine disease, and skin biopsy sampling. Some owners may wish to have clinical pathologic analytes evaluated prior to considering a more invasive procedure such as biopsy sampling. In this case the results of clinical pathology testing would have been within normal limits, so biopsy sampling would have been required to achieve a diagnosis.

Using the pathology report:

✔Read and interpret the biopsy report.

Report Diagnosis: Epidermal and follicular hyperkeratosis with absence of sebaceous glands, mild perifollicular scarring and patchy follicular atrophy.

Report Comment: The lesions and history are compatible with sebaceous adenitis of the standard poodle.

Summary of results: The report provides a definitive diagnosis in this case. The contribution made by high quality biopsy samples and a complete dermatopathology submission form including differential diagnoses cannot be overemphasized.

Comments about the Case:

The histopathologic findings supported the clinical findings, and provided a specific diagnosis of the disease process. The list of clinical differential diagnoses provided by the practitioner are useful for the pathologist as they help the pathologist envision the types of lesions present on the animal and help the pathologist appreciate the clinical lesions from the practitioners point of view. In this case, the fact that sebaceous adenitis was on the practitioner's list of differential diagnoses was useful. In the late stage of sebaceous adenitis, the sebaceous glands are missing and it can be difficult to see "what is not there". In fact, sebaceous adenitis was misdiagnosed for many years until a dermatopathologist, looking at a series of samples collected from alopecic scaly poodles, recognized what was missing (the sebaceous glands). The inflammation in the skin samples had subsided long before the samples were collected. The clinical differential diagnosis of sebaceous adenitis should serve as a reminder for any pathologist evaluating biopsy samples to double check the sebaceous glands.

Recommended Readings

Veterinary Dermatology

Mueller RS. *Dermatology for the Small Animal Practitioner*. Jackson, Teton NewMedia, 2000.

Scott DW, Miller Jr. WH, Griffin CE. *Muller & Kirk's Small Animal Dermatology*. 6th ed. Philadelphia, WB Saunders Co., 2001.

Veterinary Dermatopathology

Ginn PE, Mansell JEKL, Rakich PM. The skin and appendages, in *Pathology of Domestic Animals*, 5th ed. Maxie MG and Slocombe RF eds., Oxford, Elsevier Ltd., In Press..

Gross TL, Ihrke PJ, Walder EJ, Affolter VK. *Skin Diseases of the Dog and Cat: Clinical and Histopathologic Diagnosis*, 2nd ed. Oxford, Blackwell Science, 2005.

Hargis AM, Ginn PE. The Integument, in *Pathologic Basis of Veterinary Disease*, 4th ed. McGavin MD, Zachary JF, eds. St. Louis, Elsevier Mosby, 2006.

Yager JA, Wilcock BP. *Color Atlas and Text of Surgical Pathology of the Dog and Cat. Dermatopathology and Skin Tumors*. Spain, Mosby-Year Book Europe Ltd., 1994.

Veterinary Neoplasia

Goldschmidt MH, Shofer FS. *Skin Tumors of the Dog & Cat*. New York, Pergamon Press Ltd., 1992.

Goldschmidt MH, Dunstan RW, Stannard AA, von Tscharner C, Walder EJ, Yager JA. *Histological Classification of Epithelial and Melanocytic Tumors of the Skin of Domestic Animals*. 2nd series, Vol. III. Washington, D.C., Armed Forces Institute of Pathology, 1998.

Goldschmidt MH, Hendrick MJ. Tumors of the Skin and Soft Tissues, in *Tumors in Domestic Animals*, 4th ed. DJ Mueten, ed., Ames, Iowa State University Press, 45-117, 2002.

Hendrick, MJ, Mahaffey FA, Moore FM, Vos JH, Walder EJ. *Histological Classification of Mesenchymal Tumors of Skin and Soft Tissues of Domestic Animals*. 2nd series, Vol. II. Washington, D.C., Armed Forces Institute of Pathology, 1998.

Human Dermatopathology

Ackerman, AB. *Histologic Diagnosis of Inflammatory Skin Diseases: An Algorithmic Method Based on Pattern Analysis*. 2nd ed. Baltimore, Williams and Wilkins, 1997.

Elder D, Elenitsas R, Jaworsky C, Johnson Jr. B (eds). *Lever's Histopathology of the Skin*. 8th ed. Philadelphia, Lippincott-Raven, 1997.

Hood AF, Kwan TH, Burnes DC, Mihm Jr. MC. *Primer of Dermatopathology*. Boston, Little Brown and Company, 1984.

Index

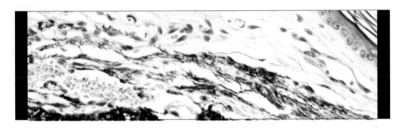